THE HISTORY OF TAMIL LANGUAGE SCIENCE AND THE WESTERN CONNECTION 1543-1875

THE HISTORY OF TAMIL LANGUAGE SCIENCE AND THE WESTERN CONNECTION 1543-1875

S. Jeyaseela Stephen

KALPAZ

First Published, 2019

Published by

Kalpaz Publications
C-30, Satyawati Nagar,
Delhi – 110052
Ph.: 011-47034999, 9811692060
www.kalpazpublications.com
E-mail: kalpaz@hotmail.com

Printed at: G. Print Process, Delhi

Cataloging in Publication Data—DK
Courtesy: D.K. Agencies (P) Ltd. <docinfo@dkagencies.com>

Stephen, S. Jeyaseela, author.
The history of Tamil language science and the western connection, 1543-1875 / S. Jeyaseela Stephen.
pages cm
Includes bibliographical references.
ISBN 9789353240769

1. Tamil language—Grammar—Study and teaching—Europe—History. 2. Tamil language—Dictionaries—Study and teaching—Europe—History. 3. Tamil language—Study and teaching—Foreign speakers—History. I. Title.

LCC PL4751.S74 2019 | DDC 494.8117 23

Contents

Illustrations

Preface

Over the three decades, as I have been sifting through the pages of Tamil history in the early modern world, I found that the language, literature, linguistics and culture had been sparkling impressively through the interaction and exchange of global contacts. Having contributed from time to time to the study of maritime trade and commerce, migration and movement of people, the transfer of ideas and scientific knowledge, I was fully convinced that a wide gap exists in the context of a consolidated study of Tamil language science in particular that evolved around the world through the efforts of the missionaries. Hence I began to work and the present work substantially fills that gap by providing an uncompromising study of primary sources in various European languages.

It is my great pleasure to acknowledge the funding bodies—the Fundação Calouste Gulbenkian, Lisboã, Portugal and Fondation Maison des Sciences de l'Homme, Paris that supported my research at different stages. Therefore, I extend my sincere and warm thanks. I also wish to express my gratitude to Biblioteca Nacional de Lisboã, Arquivo Historico Ultramarino, Biblioteca da Ajuda, Instituto Arquivo Nacionais/ Torre do Tombo, Lisboã; Biblioteca de Sociedade de Geografia de Lisboa; Archiv der Franckesche Stiftungen, Halle; Bayerische Staatsbibliothek, Munich; Staats-und Universitats bibliotheek Hamburg Carl von Ossietzy, Hamburg; School of Oriental and African Studies Library, London University; British Library, London; Bibliothèque Royale Albertine, Brussles; Det Kongelige Bibliotek and Rigsarkivet in Kobenhavn; Bibliothèque Asiatique des missions étrangères de Paris, Bibliothèque Nationale, Paris; Archives Françaises de la Compagnie de Jésus, Vanves; Centre des Archives d'Outre–Mer, Aix–en–Provence and Société Archeologique de Sens; Bibliotheca Nazionale Centrale, Vittorio Emmanule II, Roma; Biblioteca Apostolica Vaticana, Città del Vaticano; John Rylands University Library, Manchester; St. John's Abbey and University College Ville, Minnesota, USA; Library of Institut Français de Pondichéry, Historical Archives of Goa, Panaji; State Central Library in Goa; Madurai Province Jesuit Archives, Shenbaganur; Archives of the Gurukul Lutheran Theological College and Research

Institute, Chennai, for granting permission to consult and use the resources available. Finally I remain grateful to Sri. Amit Garg, Gyan Books for undertaking the publication.

The aim of this book is very modest, to examine and analyze the linguistic contact of the Europeans with Tamil during the mid-sixteenth and mid-nineteenth centuries. It intends to highlight the contribution of the Catholic and Protestant missionaries in a holistic approach towards the development of Tamil in its multifaceted approach in grammar, linguistics and dictionary-making. I am very hopeful that this book will be of abiding interest to everyone interested in the global confluence of Tamil language, grammar, lexicon and missionary linguistics.

Abbreviations

AFSt	Archiv der Franckeschen Stiftungen, Halle
AGLTC	Archives of the Gurukul Lutheran Theological College and Research Institute, Chennai
AHU	Arquivo Historico Ultramarino, Lisboã
AMEP	Archives de la Société des missions étrangères de Paris
AFCJ	Archives Françaises de la Compagnie de Jésus, Vanves
ARSI	Archivum Romanum Societatis Iesu, Roma
BA	Biblioteca da Adjuda, Lisboã
BAMEP	Bibliothèque Asiatique des missions étrangères de Paris
BAV	Biblioteca Apostolica Vaticana, Città del Vaticano
BL	British Library, London
BNCR-VM	Biblioteca Nazionale Centrale di Roma, Vittorio Emmanule II
BNF	Bibliothèque Nationale de France
BNL	Biblioteca Nacional de Lisboã
BNP	Bibliothèque Nationale, Paris
BSGL	Biblioteca de Sociedade de Geografia de Lisboa
CAOM	Centre des Archives d'Outre–Mer, Aix–en–Provence
DI	Documenta Indica
HAG	Historical Archives of Goa, Panaji
HB	Hallesche Berichte
IANTT	Instituto Arquivo Nacionais/ Torre do Tombo, Lisboã
IFP	Institut Français de Pondichéry
JRUML	John Rylands University of Manchester Library
KB	Det Kongelige Bibliotek, Kobenhavn
MPJA	Madurai Province Jesuit Archives, Shenbaganur
NAFR	Nouvelles acquisitions françaises

NHB	Neue Hallesche Berichte, Neuere Geschichte der Evangelischen
OIOC	Oriental and India Office Collection
RA	Rigsarkivet, Kobenhavn
SAS	Société Archeologique de Sens
SCL	State Central Library, Goa
SOASL	School of Oriental and African Studies Library, London

1

Introduction

Tamil history is a record of progress of accumulating knowledge and increasing wisdom, of continual advancement from lower to a higher platform of intelligence and well–being. Each generation had passed on to the next the treasures which it inherited, from Sangam age to the age of the advent of the Portuguese in the sixteenth century beneficially modified by its own experience. The commercial expansion of the East India Companies on the Tamil coast in the early modern age has witnessed rapid progress beyond all precedent. The contacts and the dialogue developed from the Tamil Coast to the Atlantic region and vice versa for the first time had resulted in global fluidity and gained a different momentum in diverse fields including exchange of ideas and knowledge, gaining new experience of Tamil language and culture. It is well known in the middle ages that Tamils had developed a tradition of accommodating diversity through public discourse and debate in Asia in the age of commerce with the Arabs and the Chinese who travelled from Aden to Canton and vice versa. This tolerance of diversity has been explicitly defended by strong arguments in favour of diversity, the richness of variation including fulsome praise of the need to interact with each other, in mutual respect through dialogue. In the same vein, mutual exchange and dialogue had continued and developed well even after the advent of Christianity to the Tamil shores.

Studies in Indo-European history or Euro–Tamil contacts mainly describe the narrative of conversion, missionaries and converts, and the whole range of issues concerning the interconnection between missionaries. Attempts have been made to situate the local milieu in the process of conversion to Christianity by focusing upon the intricate ways in which the converts identified and shaped their spiritual and material spheres of life.[1] Again there is heavy representation for the past four decades of overseas trade of the East India Companies which had developed and scholars have poured out a large corpus of literature of maritime history connections between Tamil Coast and Europe.[2] Recently a work on the scientific contribution between Tamils and Europeans had been published.[3] But there is very little work done on the missionary–scholars and their contribution to Tamil studies in depth. The earlier European studies of the missionary linguists were not comprehensive at all and the only exception is that of Joseph Dahlmann (1861–1930), the Jesuit titled '*Die Sprachkunde und die Missionen: Ein Beitrag zur Charakteristik der ältern katholischen Missionsthätigkeit (1500– 1800)*', appeared in 1891 which is also a century old. Some of the existing works are outdated since new manuscript material in several repositories of Europe had been discovered, identified but yet remain unused.[4] Under these circumstances, it is high time for us to think and open up a new road and draw the map in the field of Tamil language, literature, grammar, dictionaries and linguistics because multiple histories and processes have gone into the making. Hence the present work aims to explore how Tamil language, grammar, dictionary and linguistics had ushered into modernity and it is argued that the European encounter is though a significant part there was no scope of the Western style for dominating, restructuring, and having authority in spite of the gradual movement from the pre-colonial to colonial age.

The design of the study has been made into two parts. The first part analyzes how Tamil language and grammar travelled and progressed from pre-modern to modern times in missionary dialogue. Chapter one introduces the subject proper and the next chapter traces the love for language and the arrival of the missionaries and learning Tamil between 1543 and 1841 in a *long duree* perspective. It traces the painful struggle to pick–up the new language by Francisco de Jassu y Xavier (1543-8), the method of learning by Anrique Anriquez (1548-67), the lights of learning by Roberto Nobili (1606-10), Antão Proença's exercise of learning and comparing Tamil with Portuguese (1653-60), Philippus Baldaeus the Dutch missionary and his views (1660-5), Bartholomäus Ziegenbalg the German missionary and his way of learning Tamil (1706-10), Louis–Noel de Bourzés the French Jesuit and his style, (1710-5), mastering Tamil by Costanzo Giuseppe Beschi the Italian (1711-20), difficulties in learning reported by Jean–Venant Bouchet and Pierre Martin, the experience of Johann Lucas Niekamp and Johann Gerhard Köenig, the need for standardization of Tamil orthography emphasized by Christoph Samuel John in 1796, Clement Bonnand the MEP missionary in Pondicherry and his opinion on learning Tamil in 1825 and finally Louis Savinien Dupuis and Louis–Marie Mousset in Pondicherry (1833-41) and their views on Tamil and comparing with Sanskrit and other languages.

It is significant to explore as to how the missionaries found out the unique features of Tamil grammar while studying Tamil language and how they disseminated the knowledge across the globe. In this connection chapter three is devoted to the study of Tamil grammar by the missionaries and how they generated multiple grammars in European languages and circulated paper manuscripts between 1549 and 1855. We find Tamil grammar in Portuguese, in Dutch, in German, grammar in Tamil composed for literary Tamil by Beschi (1730), his Thonnul Vizhlakkam, the grammar for literary Tamil in Tamil (1731), translation of the grammar for literary Tamil into Latin by Beschi (1732), Tamil grammar in Latin by Onorato de Udine (1793) and Tamil grammar composed in French and the observation of Dupuis and Mousset the MEP missionaries at Pondicherry in 1855.

Tamils had circulated Tamil grammar only in palm-leaves. The Europeans who introduced paper began to write Tamil grammar in paper and circulated manuscripts. As the Westerners were exposed to the new order of print culture at that time, some missionaries had desired eagerly to print Tamil grammar books. Therefore chapter four examines how printing Tamil grammar in Latin and English for the western audience had developed between 1716 and 1843. Ziegenbalg's Tamil grammar in Latin was published at the printing press at Halle in Germany in 1716. Christoph Theodosius Walther's Tamil grammar in Latin was printed at the printing press in the Danish colony of Tranquebar on the Tamil coast in 1739. Beschi's spoken Tamil grammar in Latin was also first printed at Tranquebar in 1739 and the second printing was carried out at Madras in 1813 and the new edition and the third printing was done at the Catholic press in Pondicherry in 1843. Tamil grammar of Johann Philipp Fabricius in English was printed at Madras in 1789. Thus the travel and the marvel of printing Tamil grammatical works which was a novel in that age had been traced.

The Catholic missionaries established a College-Generale at Virampattinam in the environs of Pondicherry in 1775 approved by the Pope. It functioned under the special care of the Holy See and the seminary was closed in 1781 due to lack of students. Later the newly reconstituted college began functioning in Pondicherry in 1792 and Mgr. Louis Charles Hebert (1763-1836) assumed charge. Native Catholic students aspired for priesthood were admitted and trained. The first two native priests were ordained in 1788. Three others were ordained between 1802 and 1823. There were nine priests ordained by 1828. Owing to some problems the seminary was closed in 1832 but reopened in 1835 and it started functioning very well. It was at that time the teaching and the curriculum deserved printed books in Tamil for the use of the seminarians and so grammars and dictionaries were composed. In this connection chapter five traces the main activities of Louis Savinien Dupuis in Pondicherry and printing multiple grammars between 1845 and 1863. It examines the Piranchu Ilakkana Noorsurukkam (Elements of French Grammar in Tamil) printed in 1845, Lathin Ilakkana Noorsurukkam (Elements of Latin Grammar in Tamil) printed in 1845, Ilakkana Surukkam (Elements of Tamil Grammar in Tamil) printed in 1845, Ilakkana Noolaathaaram (The Basics of Tamil Grammar in Tamil) for the use of students and children printed in 1849 and the Grammaire Française–Tamoule (Tamil Grammar in French) authored by L.S. Dupuis and printed in 1863. These printed grammar books found way to French Guyana to help in teaching the children of the Tamil coolie migrants which is quite interesting.

The second part of the book deals with the then newly emerging discipline of linguistics in the early modern age and places Tamil language in a broader framework of missionary discourse and difference. It analyzes how the birth of the bi–lingual and tri–lingual dictionaries in Tamil took place and the related missionary enterprises and the paper manuscripts which in a large measure came to be prepared and circulated between 1549 and 1773 amongst the missionaries. Chapter six therefore is situated against the backdrop of the early lexicographical attempt by the Portuguese traders in Malabar and the beginnings of contacts in linguistic terms in 1498-9. Then it is followed by tracing the growth and development of Portuguese–Tamil dictionaries since 1549 and Tamil–Portuguese dictionaries since 1640. It highlights the role of the Tamil–German dictionary and Tamil–Latin dictionary compiled by Ziegenbalg (1710-8), Latin–Tamil dictionary of Bourzés (1710) and his Tamil–Latin dictionary (1731), the Tamil–Latin dictionary of Beschi (1742), and his Portuguese–Latin–Tamil dictionary of 1743, the Portuguese–Tamil–Singalese dictionary of Jácome Gonçalves (1735) and of Euzebio do Rosario (1772) and also the Tamil–Latin dictionary of Johann Philipp Fabricius (1772-3). Discussion is centered at length on how the importance of these dictionary paper manuscripts had been realized and why many copyists were employed in preparing copies and paid for the work. Further the chapter also deals with the dictionaries in palm–leaves prepared and shipped from Tranquebar to Copenhagen for posterity.

Chapter seven describes printing Tamil–Portuguese dictionary of Antão Proença at Ambalakaadu in 1679, printing English–Tamil dictionary of Fabricius at Madras in 1776 and his Tamil–English dictionary at Madras in 1779 and its revised edition that appeared in print in 1809. The thesaurus of Tamil language titled Chaturagaraathi of Beschi of 1732 was printed at Madras in 1824 and it got reprinted in 1835. Four volumes of the dictionary of the Tamil and English languages compiled by Johann Peter Rottler were printed at Madras between 1834 and 1841. Hence the significance of all these printed dictionaries had been examined in detail.

Louis Savinien Dupuis the MEP missionary in Pondicherry played a key role by printing the tri–lingual and bi–lingual dictionaries from time to time for the use of the Tamil audience between 1846 and 1875. Therefore chapter eight traces his early attempts to compile a massive work of Latin–French–Tamil–English dictionary in 1841-2. However he was successful only in preparing Latin–French–Tamil dictionary in joint venture with Louis–Marie Mousset for the use of the seminary in Pondicherry and the book was printed at the Catholic press in Pondicherry in 1846. Later French–Tamil Vocabulary by Dupuis and Mousset was prepared and it got printed in 1850. These two missionaries have used the Tamil–French dictionary manuscript of 1724 of Louis–Noel de Bourzés while compiling the first volume of Tamil–French dictionary which was printed in 1855. They have also used Beschi's Chaturagaraathi of 1732 which was printed at Madras in 1824 and got reprinted in 1835 and his Tamil–French manuscript dictionary of 1744. Thus L.S. Dupuis and Mousset honestly admitted several other works which were found useful for their compilation and the list includes other dictionaries like the dictionary of Johann Peter Rottler (1834), dictionary of Joseph Knight and Levi Spaulding printed in Jaffna (1842), and Tamil–Latin dictionary of Joseph Gury printed at Nagapattinam (1852). The preface to the Tamil–French dictionary of 1855 is therefore quite revealing to the readers. Dupuis and Mousset continued their efforts and were highly successful in completing the second volume of Tamil–French dictionary which was printed in 1862. For this compilation they have used the French–Tamil manuscript dictionary of 1745 of Beschi. They have sincerely acknowledged in their letters even before the dictionary book was printed in 1866. Reprinting the French–Tamil dictionary took place in 1870. The printed dictionaries were also sold in the French overseas colonies and so the connections and the developments from local to global had been brought out in this study of language, linguistics and history.

Chapter nine draws several conclusions and remarks based on the earlier chapters. The appendices and select bibliography had been provided at the end. It must be said that the history of language sciences had been recorded by almost all countries to know about the past, since it is a natural instinct of the man. But the early modern contacts at the global level had brought radically a different view of the world, the language and the knowledge altering the older order into a very scientific and systematic study, particularly of the language of Tamil through the efforts of the European missionaries. As a result Tamil language, its history and linguistics got recorded widely and it did not suffer from defects such as a non–critical approach, the study of a limited area and further by focusing mainly on religious and moral interpretations and neglecting the natural causes. The reader would find in this book how Tamil tradition and transmission grew by overthrowing the barriers.

Notes

1. Georg Schurhammer, *St. Francis Xavier: His Life, His Times*, 2 vols., Rome, 1977; Dennis D. Hudson, *Protestant Origins in India: Tamil Evangelical Christians, 1706-1835*, Surrey, 2000; Arno Lehmann, *Es Begann in Tranquebar: Die Geschichte der Ersten Evangelischen Kirche in Indien*, Berlin, 1955; Anders Norgaard, *Mission U.Obrigkeit: Die Danisch–Hallische Mission in Tranquebar*, 1706-1845, Gutersloh, 1988; A. Sauliere, *His Star in the East* (Revised and reedited by S. Rajamanickam), Anand, 1995; Ines G. Županov, *Disputed Mission: Jesuit Experiments and Brahmanical Knowledge in Seventeenth Century India*, New Delhi, 1999; Ines G. Županov, Missionary Topics; *The Catholic Frontiers in India, 16th–17th Centuries*, Ann Arbor, 2005; S. Jeyaseela Stephen, *Portuguese in the Tamil Coast: Historical Explorations in Commerce and Culture, 1507-1749*, Pondicherry, 1998; S. Jeyaseela Stephen, *Caste, Catholic Christianity and the Language of Conversion: Social Change and Cultural Translation in Tamil Country, 1519-1774*, New Delhi, 2008. Daniel Jeyaraj, *Bartholomäus Ziegenbalg the Father of Modern Protestant Mission: An Indian Assessment*, Delhi, 2006.
2. Tapan Raychaudhri, *Jan Company in Coromandel, 1605–1690: A Study in the Interrelations of European Commerce and Traditional Economies*, The Hague, 1962; Sinnapah Arasaratnam, *Merchants, Companies and Commerce on the Coromandel Coast, 1650-1740*, Delhi, 1980; *S. Jeyaseela Stephen, Expanding Portuguese Empire and the Tamil Economy: Sixteenth–Eighteenth Centuries*, New Delhi, 2009; S. Jeyaseela Stephen, *Oceanscapes: Tamil Textiles in the Early Modern World*, Delhi, 2014.
3. S. Jeyaseela Stephen, *A Meeting of the Minds: European and Tamil Encounters in Modern Sciences, 1507-1857*, Primus, Delhi, 2016.
4. K. Meenakshisundaram, *The Contribution of European Scholars to Tamil*, Madras, 1974; C.S. Mohanavelu, *German Tamilology: German Contributions to Tamil Language, Literature and Culture during the Period 1706-1945*, Madras, 1993; S. Rajamanickam, *The First Oriental Scholar*, Palayamkottai, 1972; S. Rajamanickam, *Viramamunivar: Thondum Pulamaiyum*, Madras, 1996; V.M. Gnanapragasam, *Contribution of Fr. Beschi to Tamil*, Ph.D. Dissertation, University of Madras, 1965.

2

The Love for Language: Arrival of the Missionaries and Learning Tamil 1543-1841

Portuguese traders were the pioneers who reached the Tamil coast for the first time in 1507. Earlier when they landed on the Malabar Coast in 1498 and during the subsequent years, they came to know the availability of pearls on the Tamil coast. They also learnt about the abundant availability of rice that was imported in large quantities to Malabar from the Tamil coast in exchange for pepper and spices. When the Portuguese arrived in Melaka in 1511 they came to know about the various types of textiles exported from the Tamil coast and the then flourishing cloth trade with Southeast Asia.[1] Francisco de Almeida, the first Portuguese Viceroy of Goa sent a team of four persons from Cochin to Mylapore to get more information about the tomb of Apostle Saint Thomas in 1507 and two of them died while the remaining two had returned with a report to be sent to the King of Portugal.[2] The commercial and the missionary activities of the Portuguese on the Tamil coast went hand in hand and the empire got expanded penetrating into several fields and all walks of life.[3]

Portuguese missionaries came to the Tamil coast and they preached Gospel and undertook conversion of the Hindus to Christianity. Therefore, they were compelled to learn Tamil for the exclusive purpose of communication. When the missionaries began to learn the language they freely expressed their opinion and the ways and means adopted to learn Tamil. They wrote all these details in Latin, Portuguese, Italian, French and other languages as they hailed from different parts of Europe. Their experience and the difficulties encountered were narrated in their letters sent to Europe. These men became the early ambassadors of Tamil language and literature to the Europe in the early modern world.

Followed by Catholic missionaries the Protestants had come to Tamil coast and they ventured to learn Tamil for evangelization purposes. They wrote extensively in German, Danish and English mentioning how they learnt the foreign language of Tamil in the absence of learning tools works to teach the Christian doctrine and to address the Tamils for conversion to Christianity. In this chapter we shall examine the views of the missionaries and their opinions on Tamil language, and the method adopted for learning and how they experimented and contributed to the growth and spread of Tamil language from the local to the global level.

The Painful Struggle to Pick-up the New Language by Francisco de Jassu y Xavier, 1543-8

Francisco de Jassu y Xavier (1506-52) the Jesuit missionary left Lisbon on 7 April 1541 and arrived at Goa on 6 May 1542. In the end of September 1542 he left Goa and landed at Manapadu the coastal village on the Tamil coast in October 1542. Then he proceeded to Tuticorin and later moved over to Kanya Kumari (Cape Comorin). In mid-February 1544 he was again in Tamil country particularly at the pearl fishery coast during October–November 1544. Later he was found at Mannar and Travancore, but between April and September 1545 he stayed at Mylapore, the Portuguese settlement. Alphonso Cypriani another Jesuit landed on the Tamil coast in 1543, but Xavier was back on the Tamil coast during January-February 1548 and stayed till September-October 1548.[4]

It is known from the missionary records that conversion of the parava fishermen to Catholicism had taken place on the Tamil coast in 1536 and the natives took baptism, six years before Xavier landed. The Franciscan missionaries had composed and taught some prayers written in Tamil. Xavier desired to teach the tenets of Christianity and he aimed at acquiring some skill over spoken Tamil. Hence he composed a sermon in Spanish explaining the rudiments of the faith what it meant to be a Christian. It was translated into Tamil but written in Roman script. Having learnt by-heart, he made sure that it could be understood by Tamils and he delivered the sermon. Xavier decided to move constantly from place to place to do conversion and it may be important at this juncture to raise a question as to how he found adequate time to learn Tamil? He was very keen to teach catechism, the acts of faith, besides the teachings from the Bible. Xavier felt therefore very important and necessary to learn Tamil.

Xavier had explained his methodology of work in a letter to the fellow Jesuits in Rome thus: 'As they do not understand me nor I them, their native language being Tamil and me Basque, I sought out the more literate among them and chose some who knew our language (Portuguese) as well as theirs. Then after many days and meetings and much labour, and they translated them into their language and committed them to memory. I went all over the place with a bell and collected as many children and adults I could. After I had brought them together, I taught them twice each day until after one month they had learnt the prayers'. He also added that in each of the villages he left the prayers written in their own language and ordered on how they were to be taught each day, once in the morning and again at the hour of vespers. After having done this in one village he went to another village.[5] The catechism composed by Xavier was short and it was based on Portuguese text. The single addition made by him was the name of St. Thomas after the names of St. Peter and St. Paul. Xavier also composed at least three manuals of Christian instruction in Tamil. He was the first ever to attempt to translate them from Spanish into Tamil language.[6]

To Xavier the Christian doctrine (*Doctrina Christiana*, the short catechism) composed in Tamil was simple enough. Following the medieval tradition of composing catechism he did so as there was no standard model. It may be mentioned that for the larger public in Tamil country it gave no scope for debate (which in many places Xavier says he did with the Brahmins) or expression of doubts as to the meaning or implications of what the convert accepted when he or she said 'I believe'. It involved an essentializing of the faith to a few simple tenets, which included a declaration of the belief in the true god, the commandments to be followed, the forgiveness of sins, the knowledge of divine providence, the knowledge of mortal sins as defined by the Catholic Church and the moral values.[7] Catechism marked the question-answer format. It increasingly came to be used as an enterprise for children.

Xavier was seriously learning Tamil and he was keen to produce Christian prayer books in Tamil. It is reported that three young men from Tuticorin had accompanied Xavier on the pearl fishery coast as his interpreters.[8] Xavier insisted on the translation of prayers unlike the Franciscans before him who did missionary work. It suggests his recognition that the Gospel had to be understood before it could be accepted. Apart from the fallacy of the truthfulness of translations to convey meaning, the priests also made incorrect translations. In fact Xavier in course of time made corrections to his own translations. To Francisco Mansilhas he wrote: 'In the creed, when you say *enaqu venum* instead of *venum*, you should say *Vichuvasam*. Since *venum* means 'I wish' and *vichuvasam* means 'I believe', therefore it is better to say 'I believe in God' than 'I wish in God'.[9] Xavier also recognized the universalizing power of the written word by seeing to it that copies of the prayers were kept with the help of which the new converts were to be continuously instructed. This made a novelty to participants to a literary tradition, a medium that was denied to the low castes in Tamil country at that period.

In his letter of 1545 from Mylapore, Xavier had mentioned that to what extent the Portuguese language had spread fast. He says that 'if here come to from our Society of Jesus, any foreigner who does not know Portuguese, they must first learn it, because otherwise no interpreters will understand them'.[10] The efforts of Francis Xavier to learn Tamil language had been narrated by Anrique Anriquez another Jesuit in his letter dated 31 October 1548 addressed to the College in Coimbra, Portugal.[11] Anriquez later mentioned to the fellow Jesuits in Coimbra in 1552 that he had great difficulty in correcting earlier translations which were full of errors.[12] These corrected prayers were finally printed titled Cartilha in Lisbon in Roman characters as first Tamil book.

Xavier had not laid emphasis on the written Tamil language but learnt spoken language and he had conversations with the Brahmins at various places that are revealing. He said that the Brahmins not only showed receptivity but also had an openness to listen to what he had to say and were open-ended in discussions.[13] In one such conversation Xavier reveals to the reader and says that 'the arguments which are to be employed with these ignorant people need not be a subtle as those that are found in the writings of professors who are very scholastic'.[14] In another context he reveals that a certain fear of being deceived and shows a lack of openness and trust in his dealings. When Xavier asked the Brahmins, what the commandments of their gods were, he mentions a long discussion that took place amongst themselves after which the oldest among them asked him 'what the god of the Christians orders them to do'. To such a reply Xavier sensed a 'ruse' which he responded to, by not saying anything until they had spoken.[15] It is thus very evident that Xavier did not concentrate on reading and writing Tamil.

The Method of Learning by Anrique Anriquez, 1548-67

The next earliest evidence of Tamil response to study by a Westerner appears in the writings of Anrique Anriquez (1520-1600), the Jesuit. He reached Goa on 17 September 1546 and came to Tamil coast and was at Tuticorin in 1547, at Vembar in 1548, and at Kanya Kumari in 1548. He became the superior of the Jesuit mission on the pearl fishery coast in 1549 with residence at Punnaikayal from 1550 to 1556. He then moved to Mannar and lived there during 1560-61 and finally as per records returned to

Punnaikayal and continued to live there until his death on 6 February 1600.[16] On his death, a Portuguese letter reports, Anriquez was revered as a holy man even by Muslims besides Hindus: 'At Punnaikayal they consider their oath most solemn and binding when they swear by Father Anriquez. Moreover, on the day he died all the Muslims of the neighbouring village of Patanam (Kayalpatnam) fasted and the Hindus of the neighbouring place fasted two days and closed all their shops and bazaars to express their grief over the death. So great was the respect and consideration everyone had concerning his holiness'.[17]

The Padroado missionaries were scholars, and they studied Tamil language and also culture in order to make converts. Since the proselytizers kept the records, our information is based on the serious efforts made by them to study of Tamil. Anriquez had described his early efforts to learn the Tamil language and we have references about his Tamil learning. While writing to Ignatius of Loyola, as early as in February 1548, he confessed that he did not know more than two words in Tamil and it was at this time that Xavier had just returned from Moluccas. Further he mentioned that it was only under the orders of Xavier that he began to learn, speak and read Tamil as soon as he settled in the pearl fishery coast. It was so very difficult that he despaired of ever being able to master it. He even gave up learning the Tamil language for some time. On account of this difficulty, he always preferred to use an interpreter. Fr João Villa de Conde had helped Anriquez by sending a Tamil interpreter for some time. The situation however became critical for Anriquez once when the interpreter left him. This incident forced Anriquez to learn Tamil through his own effort, sitting by, day and night making the learning of Tamil as his one and only occupation. However, visiting other mission stations entrusted to his pastoral care also facilitated the process of learning Tamil. He realized at that time and said in his letter that God was very much pleased to help him greatly to learn the Tamil language.[18]

In his letter to Ignatius of Loyola, Anriquez wrote that learning Tamil was a heroic endeavor.[19] In another letter written subsequently, Anriquez has given a detailed account of the procedure adopted by him to learn Tamil. He said that he had discovered a new method, just as in Latin to learn the conjugations. By conjugating the Tamil verbs he learnt the present tense, past tense, the futures, the infinitives, subjunctives, etc. He later learnt also the various cases of nominatives, vocatives, accusatives, genitives, datives and ablatives.[20] Thereafter he was in a position to know easily which was to be put first. He reported that through this way, much labour and time was spent to learn where and when the noun, the pronoun or the verb had to be used according to the principles of Tamil grammar. This systematic working method helped him learn Tamil within a short time so that when he wrote in their language, the Tamils were greatly astonished at his mastery.

In a subsequent letter Anriquez mentioned that he did not use any interpreter for the past four months as he had learnt to read and write good Tamil. Although he spoke and preached to the paravas in Tamil, he found its pronunciation very difficult. He said that his spoken language was very different with Portuguese accent and at times, not all the people present in the church understood him. Therefore, whenever he gave some instructions in the church, he asked someone to repeat the same words in Tamil so that all those who had assembled there could follow his language.

Anriquez has noted that the nouns in Tamils were classified as rational and irrational according to the neutral gender. All the inanimate substances and irrational beings were classified as neuter in Tamil. The vowels were grouped into two, as long and short, while the consonants were known as plosives, nasals and laterals.

Missionaries who decided to learn Tamil took the help of natives to learn the language. In a letter addressed to Ignatius Loyola, it was mentioned that the missionaries at Punnaikayal received help from Pero Luis, a native Brahmin boy for the study of Tamil who lived in the house of the missionaries.[21] It was a great help to learn the language as the boy was intelligent. He quickly identified the errors in the spoken language of the missionaries. In order to improve their own spoken language they decided to speak only Tamil among themselves and imposed punishment on those who did not adhere to this rule.[22] Anriquez was of the firm opinion that any-one who really tried hard, could learn the Tamil language within eighteen months if he used the grammar properly as his own experience had shown.[23] Anriquez had thus learnt very well to write, read and speak fluently Tamil. Yet he continued and involved himself to master Tamil. We find Pero Luis, the Brahmin was teaching Tamil even in the year 1567. Anriquez mentioned in his letter of 31 December 1567 that Frei Francisco Durão knows Tamil language went to Goa to deal with some matters of the Christians. At that time he wrote also to the provincial to see if this brother could learn some Latin so that he may be ordained. One of the deacons was very advanced in the Tamil language which he has studied for the past eight months. The other deacon was not so advanced in learning Tamil because both his abilities and health are weak. The one who is advanced in Tamil language could have been already ordained and doing confessions, if he did not have other duties than learning Tamil language.[24]

The Lights of Learning by Roberto Nobili, 1606-10

Roberto Nobili (1577-1656) left Lisbon on 28 April 1604 and landed Goa on 20 May 1605. He reached Cochin in January 1606 and left in the month of March and was at Tuticorin in May 1606. He reached Madurai in November in 1606 and worked there as a missionary and later moved over to Mylapore and died in 1656.[25]

Nobili evinced keen interest in learning Tamil and while in Tuticorin he was busy studying Tamil as early as in May 1606. Writing to Cardinal Bellarmine on 1 December 1607 Nobili remarked that he made very little progress in learning Tamil for he was sick often.[26] Most of the Catholic missionaries in the seventeenth century began to have a Tamil equivalent or Christian names or created new Christian names for them. They could not and did not use their original European names while doing missionary work. Nobili was known Thathuva pothaga swami.

Alberto Laerzio, the Jesuit Superior of the Malabar Province in 1606 evinced interest to post Roberto Nobili in Madurai. He felt that it would enable Nobili to learn Tamil and do missionary work. Laerzio mentions in his report that in the month of March 1606 when he went to visit that residence at Madurai, he left there Roberto Nobili with Fr Gonçalo Fernandes so that he might learn the language of the interior country which was pure and more refined (than on the Tamil coast) and to see whether he could find some means of opening a door for the conversion of the natives.[27] Thus Nobili had come to Madurai to begin his missionary work in 1606 and he in the beginning had evinced deep interest in philosophy also and rapidly gained mastery over Tamil.

Nobili himself had stated that Tamil was a beautiful language which was so elegant and abounding in words but the most difficult language to learn. He had further stated that he gave all his free time to the study of Tamil which he learnt in six months. He said that if he had not been sick, he would have learnt it even more quickly. In a letter Nobili also wrote to Gregory (his companion during his days as a youth) and described the first steps taken by him for adapting his life as an Indian ascetic to win the admiration of the Tamils and converse with them.[28] He became more convinced that in order to overcome the mistrustful attitude of the Hindus towards the Portuguese and come closer to the Tamils, it was necessary to get to know deeply their beliefs and traditions and also accept partly their culture by going so far as to live according to their way of life. The uncompromising orthodoxy of his colleagues however inflicted on him the charge of heresy and he was forced to justify himself by writing his Apologia.

It is significant to know in the words of his contemporaries, how Roberto Nobili learnt Tamil and acquired mastery over Tamil. On 20 November 1609, Alberto Laerzio, the Jesuit Provincial wrote to the Superior-General in Rome as follows: 'Nobili had learnt Tamil like the Brahmins who were the most proficient in that language. He has also read many of their works and learnt by-heart certain portions as well as many verses of their most famous poets'.[29] Fr Antonio Buccerio in his report sent in 1610 to the Superior-General mentioned that Nobili's knowledge of Tamil was particularly amazing that one would think it was his native language. It is wonderful how he had been able to master it so well within few years.[30] Fr António Vico in his letter dated 25 October 1610 wrote to the Jesuit Superior-General in Rome that he was fully convinced about the capacity of Nobili to undertake evangelical work in Madurai and he called Nobili as the ideally perfect missionary and further he highlighted the knowledge to defend the faith with Hindus. He wrote 'Shall I speak to you of Nobili's knowledge that exposes the most abstruse question of theology as if they were child's play? With that subtle talent he knows how to interest and charm others while making himself understood by the ignorant. He astonishes people with the fertile eloquence by the richness of expressions and also in respect of the variety and different idioms'.[31]

Antão Proença's Exercise of Learning and Comparing with Portuguese, 1653-60

Antão Proença (1625-1666) came to Goa in 1647 and was with the Madurai mission since 1653. He became the superior of the mission in 1655. He wrote annual letters of 1655-1656, 1659, 1660, 1662, 1663, 1664 and 1665. He died at Thottiyam on 14 December 1666.[32] Proença has left behind a much more scholarly account of his understanding of Tamil language and its linguistic nuances of pronunciation, vowels and consonants. He started learning Tamil in 1653. He pointed out that the Tamils designated vowels by a beautiful metaphor as *uyir* or the life of a word and consonants as *mey* or the body. The fusion brought about between the vowel and the consonant was called as *uyirmey* or an animated body. The vowels were named on the basis of the place and manner of articulation like gutturals, palatals and labials. The missionary had given the form, value and pronunciation of the Tamil characters of vowels.[33]

According to Proença, the explanation of these vowels presented no difficulty because they corresponded to the same in Portuguese. The Tamils however had their peculiarity and differentiated between the long and short vowels 'a' and 'I' and 'u'. With 'e' and 'o', the same figure was used for the long and short vowels. The Tamil language also had the diphthong 'ay' formed from the short 'a' and 'I', although the sound tends more toward the 'e' than 'a' and the characters (ay) and (au) were seldom used because there were only a few words which begin with these diphthongs. The sound 'au' is made up of the short 'a' and 'u' but it was not common as the first diphthong even in the medial position when the diphthong occurred in writing, the Tamils used another form, as was also the case with (ai) in the initial position.[34]

Proença further mentioned that the letter 'ka' corresponded to the Portuguese q or c. It was sometimes used as g especially when it was written singly in the middle of the word.[35] The pronunciation of the letter 'gna' was in itself very difficult. In order to explain the letter he used the word *maangaa* and pronounced it up to the letter 'n'. It was not produced with the tip of the tongue, but by touching the palate with the base of the tongue near the throat, keeping the tongue in that position without striking or pronouncing the 'g'. He said that if they strike that position of the tongue with 'a' or with any other vowel they would produce the full value of the 'gna' letter. He further added that this letter was rarely found preceding a vowel but always occurred before 'ka'.[36]

The letter 'sa' was pronounced as 'ch' but in writing sometimes it served as 'sa' and also sometimes as 'x' and at other times as 'j', for want of characters which Telugu had. Although some Tamil purists did not want or did not know how to pronounce 's' or 'j' they pronounced 'x' not exactly but as a mixture of 'x' and 's' and 'j'. All these four different sounds were, as he said, natural to the Portuguese language and it presented no difficulty for the Portuguese people.[37] The letter 'gna' corresponded to the Portuguese sound 'nh' in the words cunha or castanha. When in the middle of a word, it occurred without a vowel, it was found before 'sa' or before and other 'gna'. Antão Proença further explained that the letter 'da' was completely foreign to Portuguese pronunciation but it was absolutely necessary to understand its natural sound. He said that the sound was made by striking with the tip of the tongue turned up slightly towards a spot where the plain and smooth surface of the palate begins. Whenever this word was found doubled (da da) the force with which it was struck also doubled and it became like the sound of Portuguese 't'.[38] The letter 'zha' which the Portuguese called thick 'l' and 'zha' curved 'l' and 'ra' corresponded to the Portuguese double 'r' as in the words torre, morra, etc. He noted that the letter 'da' was never written at the beginning of a word.[39] Proença ended his explanation pointing out that the school children pronounced the vowels such as 'a' 'aa' 'I' 'ii' as 'ana' 'avana' 'ina' and 'iana' respectively.[40]

The order of the consonant letters was also so perplexing to him and contrary to the Portuguese language with its variety of n's, r's and 'l's.[41] The variety of the letters of the Tamils who, for want of enough letters, had several pronunciations for the one and the same letter as one might see in the letter 'sa' which served for 'cha', 'xa' and at other times for 'sa' or 'ja' also baffled him.[42] The word suram could be read in four ways, i.e. churam, xuram, juram and suram. The first two pronunciations meant nothing. The third meant 'fever' and the fourth 'voice'.[43] The same was also found in the letters 'sa', 'ka', 'pa' for the word kaayam may be read as cayam or gayam, the first meaning 'body' and the second 'wound'. The word thandu might be read as tandu or dandu, the first meaning 'stalk' and the second 'army'. The word paathai may be read as padey or badey while the first meant the 'road', the second 'torment'. Proença further noted that in order to teach easily to distinguish these various ways of pronouncing the letter, a new method should be used.[44] He really meant that the difficulty in pronunciation arose only because the letter 'sa' alone had four different pronunciations. He desired to insert four kinds of marks with dots, lines, or other signs to distinguish the normal letter.[45]

Philippus Baldaeus and his Views, 1660-5

Philippus Baldaeus (1632-1671), the Dutch *predikant* worked at Nagapattinam in 1660 and at Tuticorin in 1661. He aimed to learn Tamil. He wrote a short orthography of the Tamil language and said that Tamils wrote upon the leaves of the wild palm-trees with iron pencils. Their letters are distinguished into eight types such as (1) short or running letters (2) others long ones (3) some are vowels which they call letters of life because they give life to the consonants (4) some consonants which they call bodily letters, because the vowels are their found and life (5) others they call mixed letters, the same with Dutch diphthongs, composed of two vowels (6) some are called initial letters, used only in the beginning of a word (7) some only in the middle, and (8) some in the end of the word such as are used only in the end as will more clearly appear out of the annexed cuts which he included in his travelogue.[46] Baldaeus moved over to Sri Lanka in 1662 and worked as a missionary there. He added that seeing that the Tamil letters have hitherto not appeared in public print, either in the Netherlands or Germany as on 1671, it will not be amiss to allege the reasons therefore, and to show that this language is no less worth our care, than the Hebrew, Chaldean, Arabic, Persian, Samaritan and other languages.

He raised a question as to why the Tamil language had remained so long unknown to the Dutch. It was chiefly because that Tamil country was not conquered by the Dutch East India Company till the years 1661-63, from the Portuguese. He said that he was sensible to what obstacle the ignorance of the native language was to his good intentions, were well enough versed in the Dutch and Portuguese languages, yet some belonging to the more remote churches, who had been but slenderly instructed in the principles of the Christian religion by the Catholic missionaries were strangely at a loss till he got a good interpreter who being well skilled both in his mother-tongue the Tamil and the Portuguese language, did him great services for eight years together. His name was Francis de Fonseca and it was chiefly with his assistance, that the fundamental points of Christian religion, being comprehended in a small treatise, were translated out of the Portuguese into the Tamil language, which had been since printed, in 1671 at Rotterdam by John Bortstius. Francis de Fonseca the native Tamil speaker of Jaffna served Baldaeus as his Tamil teacher and helped him very much to learn Tamil since 1664.[47]

Baldaeus said that the multiplicity of business in so weighty a concern as the salvation of souls, having been so small hindrance in encompassing so difficult a language as the Tamil, he was forced to be contented with what part thereof he could attain, and those few intervals of time he had left, whereof he had given some specimen in the following pages and cuts. He could not upon this occasion pass-by in silence the considerations which might induce the Dutch to propagate the Tamil language with more care, than had been done hitherto; it being certain, that if the Christian religious worship could be firmly established in these parts (whereof the knowledge of the language was one of the chief means) it would be very instrumental to reconcile not only many of the inhabitants, but also even of the Indian princes themselves to our interest, a piece of policy well understood by Dom Manuel, the King of Portugal when he introduced the Roman Catholic religion by the help of the Jesuits in India.[48]

Bartholomäus Ziegenbalg's Way of Learning, 1706-10

Bartholomäus Ziegenbalg (1683-1719), the protestant missionary left Copenhagen in a ship that sailed on 29 November 1705 and arrived at Tranquebar the Danish settlement (after 222 days of voyage) on 9 July 1706. He worked in the Danish mission and died at Tranquebar on 23 February 1719 and buried on 24 February 1719 in the New Jerusalem Church.[49]

Ziegenbalg desired to learn Tamil and he mentioned that his daily practice of reading, hearing and speaking the Tamil language in Tranquebar enabled him to master this new language within a year.[50] According to him

Tamil the language of the black-brown people (Hindus) had long been formed into a written language. The natives wrote their books and letters with an iron stylus on palm-leaves. They possessed authors on almost all sciences and were expert in many trades and handicrafts. Most Christians in Europe suppose the Tamils to be a very barbarous people but this opinion arose from the Europeans who had been amongst them not understanding their language.

Ziegenbalg said that he must acknowledge that when he first came amongst them, and he could not imagine that their language had proper rules or that their life had the laws of civil order and took up all sorts of false ideas on their actions as if they had neither a civil or moral law. But as soon as he had gained a little acquaintance with their language and could talk to them on various subjects, he began to have a much better opinion of them and when at last he was able to read their own literary works, he found that the Tamils discussed the same philosophical texts as the savants of Europe and that they had a regular written law wherein all theological subjects were treated and demonstrated.

Ziegenbalg stated that he began learning Tamil through translators in the beginning. Mudaliappa who was speaking Portuguese had offered himself as the servant and translator of the Danish mission. Two months after his service Ziegenbalg wrote to his superiors at Halle in Germany about Mudaliappa and said that as soon as they came to anchor Tranquebar, a man of the Malabarian race came to their ship, made some inquiry about them, and asked Ziegenbalg whether he would not take him to be his servant? Ziegenbalg thereupon consulted Pluetschau his colleague, and they both thought it to be providence; and so took Mudaliappa who was of about twenty years of age into their service. Ziegenbalg also said that Mudaliappa's mother was still alive. Ziegenbalg and Pluetschau had thus set to learning Portuguese from Mudaliappa. After eight days Mudaliappa began teaching the missionaries, and he asked them in Portuguese whether he might always stay with them and go to Europe one day. They said that he would have to become a Christian and learn German (then called High-Dutch). Mudaliappa had then replied that 'as for Christian religion he would first to be instructed in the fundamental principles thereof, and get a competent knowledge of Christian way of worship'. But as for German, he fell to it immediately, beginning then and two months later he began to read and speak many things pretty well.[51]

On 3 September 1706, after two month's arrival in Tranquebar it had been stated by Ziegenbalg that he and Pluetschau had hired a Tamil tutor by name Ganapathi Vaathiyar who was a poet and well versed in all forms of Tamil diction.[52] He was seventy years old, who knew no Portuguese. He taught Tamil to them and other students in their house. It has been stated that the tutor was not able to talk to Ziegenbalg directly, but he introduced him to Tamil writing and reading. Ziegenbalg thus had no access to Tamil vocabulary and linguistic structure and he could not translate also the meanings of Tamil words and grammar.[53] Hence they planned to get this situation and got remedied. They thereafter employed Alagappan (Aleappa as found mention in records) of Vellala caste who spoke Portuguese, Danish, Dutch and German well.[54] Since 12 October 1706 Alagappan the former interpreter of the Danish East India Company in Tranquebar fluent at Tamil, German, Portuguese, Dutch and Danish developed Ziegenbalg's knowledge of Tamil, the senior teacher of seventy years who also continued to teach Ziegenbalg, was able to explain Tamil texts very well, and also wrote down new words and expressions for him, and put complex poetry into colloquial speech.[55]

In a letter to Germany dated 25 September 1706 Ziegenbalg wrote that Tamil should be taught in Europe not only because the Malabarees are a great and innumerable people but also because Europeans should thus be able to understand from their writings the mystery of their theology and philosophy out of which one might extract much that was good and sensible. Indeed he said that he must confess that his seventy years old tutor often asked such questions to make him realize that in their philosophy everything was by no means very unreasonable as Europeans in their land usually imagine about. Ziegenbalg wrote on 12 October 1706 that he had employed him as his interpreter and learned from him also the basics of Tamil grammar. Ziegenbalg had also reported that he was helped by Hans Paulsen in Tamil learning who interpreted things.[56] Benjamin Schultze (1689-1760) the protestant missionary in Tranquebar said that Ziegenbalg often confused the 'B' with 'P' and 'D' with 'T'. Ziegenbalg spoke and wrote Damul for Thamizh and Tschidamparan for Sithambaram.[57]

On 22 August 1708 Ziegenbalg wrote as to how he practiced Tamil from 8 am to 12 noon. In the presence of the copyists and a Tamil poet, he read only Tamil literary works which he had not yet read. The poet explained to him the details of the context and the history of the subject matter that he came across in some books. When he read poetry, the poet explained to him the things that were either difficult or in idioms. The copyists wrote down all the new words and phrases that were found in every literary work that he read. In the beginning Ziegenbalg had Alagappan the interpreter and so he did not need help of copyists. Later when Alagappan helped

Plutschau, the colleague of Ziegenbalg to progress in learning Tamil language, Ziegenbalg needed assistance of copyists.[58] Ziegenbalg in fact reported on 15 January 1709 that he had learnt Tamil and improved his language skill through interaction with catechist Malayappan (Pedro in Portuguese, Peter in English) who wrote down everything with his steel pen which was to be translated into Tamil.[59]

Louis-Noel de Bourzés and his Style, 1710-5

Louis-Noel de Bourzés (1673-1735), the Jesuit opted to join the Japan Jesuit mission and he reached Goa in 1704. He was sent back to Portugal by the Viceroy for having sailed without the necessary permission of the King of Portugal. He returned to Goa again with the permission of the King but could not leave for Japan since revolution broke out there. Hence he changed his mind and joined the Madurai Jesuit mission. He learnt Tamil for two years during his stay on the pearl fishery coast. He worked in the Madurai mission from 1710 to 1721. He became the Jesuit superior of mission in Madurai and served for three years. He wrote the annual letter of the Madurai mission in 1712 and later came to Pondicherry in 1714. Since 1721 he worked on the pearl fishery coast again and died on 25 February 1735.[60]

Bourzés knew Greek, Hebrew, Latin and Portuguese besides French his mother-tongue. He evinced keen interest to learn Tamil. According to him, Tamils varied a great deal in their language spelling. He noted that speaking the alphabet 'pa' was often confused with 'va' and the alphabet of 'sa' with 'ya'. Therefore he had adopted the spelling that he had considered to be better. Bourzés also said that there were differences in spoken and written forms of Tamil words such as usaram (colloquial)/uyaram (literary).[61] According to him the Tamil writing system was very imperfect. This arose inter alia from the fact that the Tamils had borrowed a large number of terms from Sanskrit, without adopting all the letters as had more wisely been done by the Telugus.[62] This shortcoming was morally beyond remedy. Bourzés mentioned that some letters were problematic in the Tamil writing system. He had tried to assist the reader in his Tamil-French dictionary which he had compiled in 1724 by putting a small dot above it. Tamils had often used this sign and they called it as 'pulli' in vernacular.[63] He added that there were two kinds of Tamil, such as senthamizh and kodunthamizh. The former was the elevated that had been used by the poets and their commentators even when they wrote in prose while the latter was the ordinary Tamil, spoken generally.[64]

Bourzés opined that Tamil language loved a multiplicity of nouns.[65] He noted the independence of Tamil language from Sanskrit. He found the relationship between Sanskrit and Latin mainly via lexical series such as the names of the zodiac and the planets, the days of the week etc. Bourzés wrote what was most surprising to him that being the relationship that he found between the number of words in Sanskrit and Latin.[66] His advice for beginners runs thus: 'one should not expect to find in the languages of India the same expressions, the same metaphors, the same ways of thought, as in European languages. Beginners often fall into the trap of wishing to fit the local language to our own, instead of accommodating themselves to the local language, its modes of thought, expressions and tone'.[67]

Mastering Tamil by Costanzo Giuseppe Beschi, 1711-20

Costanzo Gioseffo Eusebio Beschi, known as Costanzo Giuseppe Beschi (1680-1747) the Jesuit left Lisbon in 1710 in a ship for India. He joined the Madurai mission on 8 May 1711 and he was at Gurukkalpatti, at Aavur in 1711, Kamanayakkanpatti in 1715, Elakurichi and Kunampatti in 1731-34 and at Tiruchirapalli in 1740-41, and at Tuticorin and Manapadu during 1742-45. He was appointed the visitor of the Jesuit mission in 1746. He died on 4 February 1747.[68]

Beschi was very keen to master Tamil. Availing the leisure enforced upon he gave himself to the study of Tamil poetry; for nothing in the Tamil country was esteemed more than this study, wrote Louis-Noel de Bourzés in his annual letter from Kamanayakanpatti to the Jesuit Superior-General in Rome. In the space of a few months, Beschi had made progress in it than any other Jesuit missionary since the foundation of the Madurai Mission. The result was that he wrote Tamil verse most elegantly, and what was more difficult, he had read the ancient poets with much profit.[69] In 1721 Fr Nadal wrote to the Jesuits in Paris that Beschi knew Tamil to perfection and wrote beautiful verses. He was highly successful in learning both spoken/ordinary Tamil (called kodunthamizh) and literary Tamil (senthamizh). Beschi learnt Tamil under the famous poet Subramania Dikshitar and not under Subradeepa kavirayar as many scholars had mentioned.[70]

The Difficulties in Learning Reported by Jean-Venant Bouchet, Pierre Martin and Dominique de Valence

Having observed the success story of learning Tamil by the missionaries there had been also serious difficulties reported. However, the main purpose of Learning Tamil language had been neatly spelt out by the missionaries and it was for the sake of conversions to Christianity. Jean-Venant Bouchet born at Fontenay le Comte on 10 April 1655 joined the Society of Jesus on 7 October 1670. He came to serve the Madurai mission in 1689 and spent for almost twelve years. He moved from Aavur near Tiruchirapalli to Pondicherry in 1710. He was the parish-priest of Ariyankuppam in the environs of the French colony in 1713. Bouchet had said that Tamil grammar composed by a Jesuit already available was very helpful for him to learn to write Tamil.[71] Bouchet was first taught Tamil by Francisco Laynez, the procurator of the Madurai mission.[72] Bouchet wrote openly and he complained of the difficulties of beginning to learn Tamil language when he was already at an advanced age and the elements of a language which has no connection with those of which he had been used to in Europe. He stated that learning to speak Tamil particularly at a certain age was difficult because the nerves of the tongue no longer supplied enough to catch the pronunciation of certain alphabets.[73]

Pierre Martin (1665-1716) the French Jesuit joined the Carnatic mission in 1709 and he was at Pondicherry and later at Vadurgarpettai in 1713. As a procurator of the Carnatic mission he went to Rome where he died on 28 June 1716.[74] It has been admitted by Pierre Martin that he clearly lacked the ability to make inspiring speeches in Tamil.[75] As the missionaries urgently felt the need to teach the tenets of Catholicism to Tamils, they took sincere efforts in learning alphabets, speaking and writing Tamil. They desired very much to understand and speak Tamil for the sake of carrying on their day-to-day apostolic work. Dominique de Valence the French Capuchin missionary in Pondicherry described Tamil language as a harsh, crudely fashioned, unpleasant, and repelling one, especially in its pronunciation. According to him it was only a zeal for the propagation of faith which made it possible to learn this language.[76]

The Experience of Johann Lucas Niekamp and Johann Gerhard Köenig

Johann Lukas Niekamp in 1740 who wrote on the history of the German-Tamil exchange of ideas between 1706 and 1736 described that Tamil was a complete and solemn language, especially when it comes to terms of abuse. Apart from some earlier Persian and more recent European loan words, the Tamil, or as it was called in his times, the malabarian language by the Germans and Europeans, contrasts sharply from the other European and oriental languages. One might according to Niekamp even ask, how the old history of Tamil was related to the history of other languages. The word order in Tamil was said to differ considerably, when one translates a sentence literally, it seems to be wrong or disordered.[77] Niekamp summarized the key elements of the Tamils and under the headings of languages, intellectual composition of the Tamils and arts and sciences of the Tamils. He wrote that the Tamil language was ancient and likened it to Persian and European languages. However, he pointed out that unlike European languages, Tamil normally emphasized the weaker aspects of life more than the stronger one: mother and father, moon and sun, and not vice versa. At that time, it did not employ punctuation marks. Nevertheless, it was clear, accurate, powerful, and suitable for Bible translation. He said that Tamil sentences contained many infinities, participles, and gerunds. However, a single verb closed a sentence. Unlike German, Tamil had shorter, yet powerful words. It doubled certain consonants like 'k', 'p' and 't' and the sound 'h' is absent in Tamil. All vowels were linked to the consonants.[78]

Protestant missionaries were requested to learn Tamil prior to their arrival on the Tamil coast by 1784. It had been stated that missionaries will not find anyone either in Europe or on the ship who would be able to teach this Tamil language properly. To try and learn something from books and grammars or from an East India sailor, or even from a Tamil who has not learnt the rules of the language in school was unnecessary, indeed even harmful. It was felt that the missionary learner would internalize mistakes which he will find difficult or even impossible to correct later.[79]

Johann Gerhard Köenig (1728-1785) born on 19 November 1728 in the Duchy of Courland (near Kreutzburg in Polish Latvia) was trained as a pharmacist in Riga. In 1748 he took appointment in Denmark and during 1757-58 studied natural history in Uppsala, Sweden, as a student of Carl von Linne. In 1759 Köenig was appointed to the Royal Frederick's Hospital in Copenhagen as a surgeon and pharmacist, being by special permission allowed to study medicine at the University and he completed. He was appointed as surgeon at Tranquebar in the Danish settlement on the Tamil coast arrived in 1768.[80] It has been mentioned by Johann Gerard Köenig vide his letter 12 January 1771 that the language of the Tamil books had three different vulgars. One was used by the Tamils and in the literary style. The next was employed with Portuguese vulgar on the Tamil coast with three different manners with vocabularies. The third was the one that was spoken by the foreigners, spoken by the Portuguese citizens, spoken and understood by the Portuguese necessary to serve them in total. Köenig reported that under these circumstances it was very difficult to learn Tamil language.[81]

The Need for Standardization of Tamil Orthography Emphasized by Christoph Samuel John in 1796

Christoph Samuel John was born at Frobersgrun near Greiz on 11 August 1747. He studied at Halle, was ordained and arrived in Tranquebar on 13 June 1771. He worked as a missionary and died after serving more than 42 years of labour at Tranquebar on 1 September 1813, and was buried in New Jerusalem church-yard. He learnt Tamil and published an article titled, 'Several remarks on the adjustment of the orthography and pronunciation of Tamil words'. In this he discussed the problems of orthography of Tamil. According to him each European nation had its own way of spelling Tamil words. The fact that the spelling of names was not yet fixed was considered a real problem, particularly when the mapping of the Tamil country was undertaken and going on. C.S. John relates three very practical causes for such differences in the spelling of Tamil words.[82] He said that at first it was often difficult to understand the Tamils who chewed betel all day long. Secondly the Tamils were not always willing to spell the words when asked to do so. They often wondered why the Europeans wanted to know these words, anyway. Thirdly the Tamil words are not easy to pronounce in European language. While this being the opinion of C.S. John, it may be remarked that the first reason given by him may not be proper and the second may be partly true while the third one is the truth that can be accepted. It is important to note here that the difficulties increased by the significant differences in spelling among the European themselves have been noted carefully. According to C.S. John, English was the language with the most deviations, whereas the German language was the most suitable for the transliteration of Tamil words. He stated that how strange does Tritchinipoly sound? He mentions that Tirutscinapally was much more correct. Therefore we find that C.S. John had proposed a standardization of Tamil orthography. He stated that when a person wants to be safe, one should compare the spelling with what one finds in the accounts of the missionaries.[83] He further insisted that the word Malabar, in his times, the common European name for Tamil was not in use in India itself. The language is only called Tamul or strictly Tamurh.[84] In addition to these views C.S. John had also discussed at length the way the missionaries added diacritical marks to their Latin script. It is interesting to note that we find learning Tamil language had lied in the pronunciation. Tamil was spelt in European orthography and so it needed to be spelt as by the Tamils. It has been found that most of the missionaries after 1780 had tried to spell the names of the persons and places more in accordance with the tongue of the Tamils.

The difficulty of finding a suitable Roman equivalent was best exemplified by enumerating the alphabets hitherto made to represent in a Roman dress by the British Raj. It pointed out that Ziegenbalg in 1714 employed 'rl' and 'rhl'. Beschi in 1728 in his preface to Senthamizh used 'zj'. He described it as being '*quoque alind l, quod crassiori sono, reflexa omnino ad interiorem palati partem lingua pronunciatur*'. Adding in a note he said that '*in alique istius regionis parte sonat quasi j*'. The French Jesuits had used 'xh', while Ellis in 1816 had used 'zh'. Finally, the report of the sub-committee of the Madras Literary Society and Auxiliary of the Royal Asiatic Society, on writing Indian words in Roman characters finalized and gave the details of Tamil letters to be pronounced along with notes. This Tamil orthography was adopted and followed widely by many writers after 1859.[85]

The learning of Tamil language by the missionaries of Tranquebar was appreciated even by the natives like Rama Nayakkan the leaseholder of the village of Poraiyar and the adjacent area including Anandamangalam near Tranquebar the Danish settlement.[86] The missionaries who arrived in 1725 saw an elephant for the first time in Poraiyar when he arranged to see the animal. He found at that time the missionaries did not know Tamil. In March 1726, when Rama Nayakkan visited these missionaries at Tranquebar he congratulated them on their rapid progress in learning Tamil. The missionaries at that time apologized for not having the means to reciprocate a dinner since Rama Nayakkan had previously treated them to a big meal.[87]

The learning of Tamil had developed even in Europe gradually. We find the presence of a Tamil Christian by name Timothy hailing from Tranquebar who accompanied H. Plutschau to Europe in 1711, and he was asked to teach Tamil to Prof. Christian Benedict Michaelis (1680-1764), who looked after the affairs of Tranquebar mission from 1712. Timothy was also asked to teach Tamil to Heinrich Julius Elers (1667-1728) the book printer in the Orphanage Press at Halle. August Herman Francke wrote in his letter dated 31 October 1713 to his friends in Copenhagen that Timothy could stay in Halle to cultivate the Tamil language. He wished that his students in the *Collegium Orientale Theologicum* could learn Tamil and benefit very much so as to serve as the bearers of Halle pietism in India.[88] Johann Ernst Grundler (1677-1720) had sailed from Copenhagen on 17 January 1708 landed at Tranquebar, and worked as a missionary at Tranquebar.[89] He in his letter dated 15 January 1715 from Tranquebar mentioned to Prof. H.W. Schomerus that he had learnt Tamil well and progressed very much in composing works. Grundler in this letter also realized the importance of Tamil learning and wrote that Tamil language was worthy to be taught at German universities.[90]

Clement Bonnand in Pondicherry and his opinion on Learning Tamil, 1825

Clement Bonnand the missionary of the Paris Mission Society left France for India on 4 February 1824 and reached Pondicherry in 1825. He began teaching at the college in 1826 and was appointed coadjutor in 1831. He became later the Bishop of Pondicherry on 8 November 1833. According to documents Bonnand had engaged himself in learning Tamil in 1825. He felt that Tamil was a language with grotesque characters.[91] Louis Charles Hebert the Bishop of Pondicherry had given a tutor since the time of his arrival of Bonnand and so he had lessons every day. Bonnand had mentioned that he was making very little progress and he found Tamil language very ticklish.[92]

Louis Savinien Dupuis and Louis Marie Mousset in Pondicherry and their Views, 1833-41

Louis Savinien Dupuis born at Sens on 18 August 1806 aspired to become a Catholic priest and so he joined the seminary of the diocese of Sens and received religious formation. He did his studies in philosophy and theology. Cardinal de la Fare of the Archdiocese of Sens granted permission to priesthood. He was thereafter ordained a priest on 24 May 1829 by Mgr. Dupont the bishop of Samosate. After two years, he felt that he should serve as a missionary in the overseas particularly the Malabar Mission and so joined the Missions étrangères de Paris society on 25 March 1831. Dupuis who was eager to serve the mission therefore sailed from France to India in a ship that left on 4 July 1831. After sailing for about seven months in total, Dupuis reached the port of Pondicherry the French settlement in South India on 9 February 1832. The arrival has been reported by Louis Charles Herbert, the bishop in Pondicherry (1763-1836) vide his letter 12 February 1832. Dupuis stayed for six months at Pondicherry and he was engaged in learning Tamil. He was deputed for apostolic work to serve as a chaplain in the British garrison at Bangalore on 21 August 1832 by the Bishop. Hence Dupuis obeyed the orders and travelled via land and reached Bangalore.[93] Dupuis continued to learn Tamil and he mastered it so well that he was able to compose a work in Tamil against Protestantism within seven years (by 1839). He had sent a copy of it to Bishop Bonnand in Pondicherry so that it might be read and printed if judged worthy. As the main book had already been received the needed authorization for printing,

Dupuis believed that in some more time, the abridged version and the main book would be printed. Dupuis hoped this abridged version of his book would be of great use to win back the Christians who embraced Protestantism and the public. According to records we find that Dupuis was posted later to work among the Telugus. In his letter dated 18 November 1839 from Ballapuram he said that he would have to communicate in a language in which he was not at all fluent with. He mentioned that he had been busy till date looking after the mission by teaching Tamil Christians and the English. He thought that he would continue in this mission before being obliged to preach in that language. He said that the Lord was still giving him some more days of life to learn Telugu and so he would do it. Dupuis stated that he had command over French and Tamil but could not finish a sentence in English without shivering and being disturbed also by repetition.[94]

Louis Marie Mousset was another MEP missionary and he sailed from France on 22 April 1835 and arrived in Pondicherry on 31 August 1835. He wrote a letter on 16 September 1835 from Pondicherry to Jean Tesson the procurator for the missions in India in Paris wherein he stated that he learnt Sanskrit grammar during his voyage.[95] On arrival in Pondicherry he began to learn Tamil and he wrote later another letter to Tesson dated 29 November 1836 from Sarugani and mentioned that he had begun to speak Tamil well.[96] He was appointed the parish-priest of Kamanayakkanpatti in 1837 and learnt about Beschi and his Tamil writings and he was deeply inspired to master Tamil. Bonnand the Bishop of Pondicherry wrote in his letter to Langlois in Paris dated 18 February 1841 and expressed high opinion of Mousset and said that he could speak fluently and write Tamil very well.[97]

The observation on Tamil language made by Dupuis and Mousset the two MEP missionaries in the Tamil-French Dictionary published by them in 1855 is interesting to note. According to them the Tamil letters are completely different from Sanskrit in number and form. Tamil has six letters not found in Grantha: short 'ye' and short 'o' and the consonants 'izhil', 'zhal' 'irr' and 'inn'. Sanskrit has many other letters not found in Tamil. Usage has, admittedly, introduced a number of foreign characters, to express some of the Grantha letters which are lacking in this language. But the original names of the Sanskrit characters, which they retain is a proof that the other letters of the language, which belong to Tamil are not derived from Sanskrit. Is it from Malayalam that Tamil had borrowed the characters which they refer, whose only Sanskrit element was the sound and which substitute for Grantha letters. However, pure Tamil does not admit them, either in its dictionaries or in other works, and represents the letters of words borrowed from Sanskrit, by the Tamil characters nearest to them.

Lastly, Tamil has special spelling rules for Sanskrit words, either when combining together or with pure Tamil expressions—rules which serve to perpetuate the tradition of their foreign origin.[98] But the greatest proof of Tamil not being derived from Sanskrit was the complete difference that exists between the two languages in a host of areas, which given them quite distinct characteristics and are an obvious indication of separate origins. Dupuis and Mousset mention just the main features.

Sanskrit and Tamil the two languages are grammatically almost the opposite of each other. The rules of Tamil syntax are not complex and not numerous; its grammar is fairly standardized. It can be altered slightly, but there are some things which are generally invariable such as the position of the subject at the beginning of the sentence, and the main verb at the end. The order of the elements in the Tamil sentence is often the inverse of that of European languages. The reverse translation of a Tamil sentence with several clauses, that is from the end to the beginning gives a reasonably correct French or Latin sentences. Sanskrit idiom, on the other hand, is appreciably closer to that of Greek or Latin.[99]

To distinguish masculine and feminine gender, Tamil follows what nature herself has given to rational beings, or what mythology has created. All names of inanimate or irrational beings are neuter, whereas Sanskrit as well as Greek and Latin, these nouns are arbitrarily masculine, feminine or neuter. Dupuis and Mousset do not distinguish between natural and grammatical gender. In addition, Tamil has special endings for masculine, feminine and neuter in the third person singular, and one ending for the neuter in the plural, which is different from the masculine-feminine common plural. This is close to Hebrew usage. Sanskrit does not have this.[100]

For numbers, Sanskrit like Arabic follows the decimal system. Tamil is quite different. In written form of the numbers, one, ten, one hundred and one thousand are each represented by one figure. Thus one can trace the characteristics peculiar to Tamil, nevertheless, as Dupuis and Mousset said that their aim has not been to write a dissertation, but to give a brief overview of the principal bases on which for a judgment in this matter, they have confined themselves to the main features which they have indicated.[101]

The main languages which come from Tamil are Telugu, Kannada and Malayalam. These three, although they have different scripts from Tamil are basically very similar to it. Many words differ only in their inflections. The grammar is essentially the same, so that generally speaking, one can make a word-for-word translation from one to another of these languages, following the order of the words, without detriment to the clarity or the beauty of the sentence. Like Tamil these three languages have no relative who or which; they have the same tenses as Tamil a similar negative verb for each verb tense, for all persons like in the past tense in English, except that English varies in the second person singular. The forms of some Malayalam letters are similar to those in Tamil, but the scripts of both languages are entirely different from those of Telugu and Kannada, which in turn share a similarity in nearly all of their characters. As the order of the alphabet, whether Telugu, Kannada or Malayalam, is closer to Sanskrit, and these three languages have all the letters necessary to replace the Grantha characters, whereas, fundamentally they have direct relationships with Tamil, it is clear that their origin is later than this language and even probably, later than the Brahmin migration. Tamil is best known to the French as Malabar and Adriano Balby in his dictionary of geography based on peace treaties and the most recent discoveries printed in Paris in 1833 called the descendants of Tamil the Malabar family.[102]

As this Tamil language has only one declension, one conjugation, with few tenses, indeclinable adjective, neuter nouns for all inanimate or irrational beings, and few syntactic rules, one can say that it is not very difficult, especially ordinary Tamil, in which one can easily express one's thoughts as soon as one knows a few thousand words. The inverse order of sentences and the pronunciation of retroflex consonants sometimes prove difficult for Europeans. Nonetheless once one has got used to it, the grammatical structure becomes easy. The language of poetry presents more difficulties because if its laconicism, its ellipses and above all because of the polysemous nature of the words it uses.

According to Dupuis and Mousset Tamil, especially the elevated language is rich in compounds. Some are formed by joining Sanskrit word, others pure Tamil words, others again by joining one to the other. In the composition of these words, the language followed fairly regular rules, such that the words are rarely distorted, as are compounds or derivatives in many languages. This is why there are few languages as pure as Tamil in this respect.[103]

We have so far observed that how various missionaries have learnt Tamil during a long span of around 300 years. It may be mentioned that these missionaries lacked the usual means to learn a foreign language. Transliteration of Tamil characters into Latin characters was done in the course of learning. With the help of native speakers the missionaries had improved the Tamil language skill. Progress in learning Tamil was negligible in the beginning to the missionaries. Tamil was studied by all the European missionaries as they desired to communicate in local language. They wished to compose prayers in Tamil, develop Christian literary pieces and further aimed to translate the Bible and publish religious booklets. As they had practical reasons, to address the gatherings in Tamil and so both the spoken and written tongue they paid attention to master the Tamil language.

To conclude, the honour of truth should not be passed over in silence. There was no ban on the study of the pagan language of Tamil. It was studied by the pious missionaries for their own sake and it did not attract the attention of the Holy inquisition as heresy neither in Europe nor in India. The purpose of the missionaries was to reach the goals defined by the religious orders that is, to refute the Hindu doctrine. They did not learn the language out of appreciation for Tamil culture.

Notes

1. S. Jeyaseela Stephen, *Portuguese in the Tamil Coast : Historical Explorations in Commerce and Culture, 1507-1749,* Pondicherry, 1998, p. 7.
2. S. Jeyaseela Stephen, *Caste, Catholic Christianity and the Language of Conversion: Social Change and Cultural Translation in Tamil Country, 1519-1774*, New Delhi, 2008, p. 45.
3. S. Jeyaseela Stephen, *Expanding Portuguese Empire and the Tamil Economy: Sixteenth-Eighteenth Centuries*, New Delhi, 2009.
4. George Schurhammer, *St. Francis Xavier: His Life, His Times*, 2 vols., Rome, 1977.
5. M. Joseph Costelloe, trs., *The Letters and Instructions of Francis Xavier*, Anand, 1993, Document no. 48, pp. 117-8.
6. Ibid., Document no. 14, pp. 41-5.
7. Ibid.
8. Georg Schurhammer, *Epistolae S. Francisci Xaverii alia que eius Scripta*, Roma, 1944, vol. I, pp. 127, 245.
9. M. Joseph Costello, p. 78, Document no. 24, see the letter of Xavier written from Manapadu to Francisco Mansilhas in Punniakayal on 27 March 1544.
10. S. Jeyaseela Stephen, Portuguese in the Tamil Coast, pp. 400-2, see the translation of the letter of Xavier written from Mylapore on 8 May 1545.
11. Jose Wicki and J. Gomes, *Documenta Indica* (hereafter DI), 18 vols., Roma, 1948-88.
12. António da Silva Rego, *Documentação para a Historia das Missões do Padroado Portugues do Oriente*, 12 vols., Lisboã, 1947-58, vol. V, document no. 47, p. 225, see the letter of Anrique Anriquez to the Confraternity of Coimbra written from Cochin dated 17 November 1552.
13. M. Joseph Costelloe, p. 70.
14. Ibid., p. 71.

15. Ibid., p. 70.
16. S. Jeyaseela Stephen, *Portuguese-Tamil Grammar: Modernization and Democratization of Tamil*, Pondicherry, 2001, p. 21.
17. Jose Wicki, 'Life of P. Henrique Henriques S.J. (1520-1600)', *Studia Missionalia*, vol. XIII, 1963, pp. 157-160.
18. DI, vol. I, p. 279.
19. Ibid., vol. I, pp. 276–300.
20. Ibid., vol. I, pp. 285-6.
21. Ibid., vol. I, p. 279.
22. Jose Wicki, Life of P. Henrique Henriques, p. 157.
23. DI, vol. III, pp. 97-8, 601.
24. Archivum Romanum Societatis Iesu (hereafter ARSI), Roma, MS Goa, vol. 8 III, fls. 629r–33v, see the letter of Anrique Anriquez to the members in Portugal, written from Punnaikayal dated 24 December 1567. See also, letter of Anrique Anriquez to Francis Borgia the Superior-General of the Society of Jesus in Rome dated 31 December 1567 ARSI, Mss Goa, vol. 8, III, fls. 667r-68v, DI, vol.VII, pp. 437-443.
25. S. Rajamanickam, *The First Oriental Scholar*, Palayamkottai, 1972, pp. 12 and 17.
26. Ibid., p. 18.
27. Ibid., p. 19.
28. Biblioteca Apostolica Vaticana (hereafter BAV), Città del Vaticano, Bon Compagai Section E6, MS. Lettre di Roberto Nobili, Missionaria Gesuita nel Madure nellee Indie orientali dal anno 1606, fls.1-12.
29. ARSI, Roma, MS Goa, vol. 54a, fls.1-16, fls.17-55.
30. ARSI, Roma, MS Goa, vol. 54a, fls.76-92.
31. ARSI, Goa, vol. 48, fl.178, see the letter of Antonio Vico dated 25 October 1610.
32. S. Jeyaseela Stephen, *Letters of the Portuguese Jesuits from Tamil Countryside, 1666-1688*, Pondicherry, 2001, p. 378.
33. BAV, Borgiano Indiano 12, Vocabulario Tamulico -Lusitano, pp.14-5.
34. Ibid., p.15.
35. Ibid.
36. Ibid.
37. Ibid., p.16.
38. Ibid.
39. Ibid., p.17.
40. Ibid., p.18.
41. Ibid., p.12.
42. Ibid.
43. Ibid., p.13.
44. Ibid.

45. Ibid.
46. Philippus, Baldaeus, *A True and Exact Description of the Most Celebrated East India Coasts of Malabar and Coromandel as well as of the Isle of Ceylon with their Adjacent Kingdom and Provinces*, London,1703, rpt., New Delhi, 2000, vol. III, p. 664.
47. *Philippus Baldaeus, Nauwkeurige beschryvinge van Malabar en Choromandel, derhalver aangrenzende ryken, en het machtige eyland Ceylon, Nevens een omstandige en grondigh doorzochte ontdekking en wederlegginge van de afgoderye der Oost-Indische heydenen*, 3 vols., Amsterdam, 1672, pp. 191-4.
48. Letter of Philippus Baldaeus from Geervliet dated 21 July 1671 in Philippus Baldaeus, A True and Exact Description, p. 663.
49. Johann Ferdinand Fenger, *History of the Tranquebar Mission: Worked out from Original Papers*, trs., Emil Francke, Tranquebar, 1863, p. 312.
50. Archiv der Franckesche Stiftungen (hereafter AFSt), Halle, MSS Letters: M B 75/76, fl.102.
51. Bartholomäus Ziegenbalg, *Propagation of the Gospel in the East: Part III: Being a Collection of Letters from the Protestant Missionaries, and other worthy Persons in the East-Indies, & c. Relating to the Mission; the Means of Promoting it; and the Success it hath pleased GOD to give to the Endeavours used hitherto, for Propagating True Christianity among the Heathen in those Parts, but chiefly on the coast of Coromandel. With a map of the East-Indies, Published by the direction of the Society for Promoting Christian Knowledge, London, Printed and Sold by J. Downing, in Bartholomew-Close near West-Smithfield*, 1718, pp. 141-5; D. Dennis Hudson, *Protestant Origins in India: Tamil Evangelical Christians, 1706-1835*, Grand Rampids, 2000, p. 12.
52. Johann Ferdinand Fenger, History of the Tranquebar, p. 153.
53. Arno Lehmann, *Es Begaan in Tranquebar : Die Geschite der ersten evangelischen Kirche in Indien*, Berlin, 1956, p. 21.
54. Ibid., pp. 23-4.
55. Ibid., p. 24.
56. Wilhelm Germann, *Ziegenbalg und Plutschau: Die Grundungsjahre der Trankebarschen Mission: Ein Beitrag zur Geschite des Pietismus nach handscriftlichen Quellen und altesten Drucken*, 2 vols., Erlangen, 1868, vol. I, p. 65.
57. *Der Königl. Dänischen Missionarien aus Ost-Indien eingesandter ausführlichen Berichten, Von dem Werck ihrs Ams unter den Heyden, angerichteten Schulen und Gemeinen, ereigneten Hindernissen und schweren Umstanden; Beschaffenheit des Malabarischen Heydenthums, gepflogenen brieflicher Correspondentz und mundlchen Unterredungen mit selbigen heyden*, (hereafter HB), Teil 1–9, (Continuationen 1–108) Waiserihaus, Halle, 1710–1772, vol. III, continuation 25, p. 159.
58. HB, vol. I, continuation 6, pp. 282-3.
59. Bartholomäus Ziegenbalg, Propagation of the Gospel in the East, part III, pp. 146-7.

60. Leon Besse, *Liste Alphabetiques des Missionaire du Carnatic de la Compagnie de Jesus au XIIIe Siecle*, Pondichéry, 1918; *Lettres édifiantes et curieuses, écrites des missions étrangères par quelques missionaires de la Compagnie de Jésus* (hereafter Lettres édifiantes et curieuses), 34 vols., Paris, 1702-1777, vol. XIII, pp. 212-6, see the letter of Louis-Noel de Bourzes to the Jesuit Superior General in 1712.
61. John Rylands University of Manchester Library (hereafter JRUML), Tamil Manuscripts, no. 6, see preface, para 5.
62. JRUML, Ibid., preface, para 6, see for Pa, pa. Ba, bha.
63. Ibid., para 7.
64. Ibid., para 8.
65. Ibid., para 34.
66. Lettres édifiantes et curieuses, vol. XIII, see the letter of Louis-Noel de Bourzes to Souciet dated 23 March 1719.
67. JRUML, Tamil Manuscripts, no. 6, fl. 3b, para 12.
68. S. Rajamanickam, The First Oriental Scholar, p. 268. The baptismal name is Costanzo Gioseffo Eusebio and he was baptized on 13 November 1680. See, Historical Archives of the Parish of Castiglione delle Stiviere, *Battesimi*, vol.7, ab anno 1667 ad 1683. Beschi had signed always his name Costanzo Giuseppe Beschi (Constantius Joseph Beschi) adopting slightly the Latinized form.
69. Leon Besse, *Father Beschi: His Times and his Writings*, Trichnopoly, 1918, pp. 66-67.
70. Su. Innasi, *Thembavani Thiran*, Chennai, 1994, pp.12-3; Muttusami Pillei, *A Brief Sketch of the Life and Writings of Fr. C.J. Beschi*, Madras, 1840.
71. Lettres édifiantes et curieuses, vol. XV, p. 266.
72. Ibid., vol. XV, p. 226.
73. Ibid., vol. XV, p. 266-7.
74. S. Rajamanickam, The First Oriental Scholar, p. 272.
75. Lettres édifiantes et curieuses, vol. V, p. 94, see the letter of Père Martin to Père Le Gobien written from Kamanayakkanpatti dated 1 June 1700.
76. Bibliothèque Nationale de Paris (hereafter BNP), MSS Nouvelles acquisitions françaises (hereafter NAFR), no. 2627, The text runs as follows: 'La langue Tamoul est rude, grossière, désagréable, et rebutante surtout dans la prononciation. Il n'y a que le zèle de la propagation de la foi dans les missionnaires qui soit capable de porter de l'apprendre'.
77. Johann Lucas Niekamp, *Kurzgefaßte oder historischer Auszug der Evangelischen Missionsberichte aus Ost-Indien von dem Jahr 1705 bis zu Ende des Jahres 1736 mit zwey dazu nöthigen Landkarten und einer Vorrede Herrn Gotthilf August Franckens*, Halle, 1740, pp. 50-4.
78. Ibid., pp. 80-134.
79. Rigsarkivet (hereafter RA), Kobenhavn, Missionskollegiet og Direktionen for Vajsenhuset (hereafter Miskolark), 9G, Memoria for new missionaries from Christoph Samuel John and Johann Peter Rottler from Tranquebar dated 20 January 1784.

80. Johan Ferdinand Fenger, History of the Tranquebar, p. 316.
81. AFSt, I B, 73, Letter of Johann Gerhard Köenig dated 12 January 1771.
82. Georg Christian Knapp, et.al., *Neuere Geschichte der evangelischen Missions-Anstalten zu Bekehrung der Heiden in Ostindien aus den eigenhändigen Aufsätzen und Briefen der Missionarien erausgegeben* (hereafter NHB), Teil 1–8, (Stück 1–95) Waisenhaus, Halle, 1770-8/ 1795-1848, NHB, 4, 1796, pp. 725-35.
83. Ibid., 4, 1796, p. 726.
84. Ibid., 4, 1796, p. 725.
85. *The Report of the Sub-committee of the Madras Literary Society and Auxiliary of the Royal Asiatic Society, on Writing Indian Words in Roman Characters*, Fort St. George Gazette Press, Madras, 1859, pp. 21-2, 31-2.
86. AFSt, M I B2:7, p. 8; HB, vol. 2, no.20, p. 472.
87. HB, vol. 2, no. 22, p. 864.
88. RA, Ledr. 389.2.
89. Grundler was born at Weissensee on 7 April 1677, studied at Halle and was ordained at Copenhagen in 1708. He died at Tranquebar on 19 March 1720 and buried the next day in the New Jerusalem Church. See, Johan Ferdinand Fenger, History of the Tranquebar, p. 312.
90. HB, vol. I, Continuation 8, 1715, pp. 505–55.
91. Archives de la Société des Missions Etrangères de Paris (hereafter AMEP), Correspondance Pondichéry-Paris, Lettres, vol. 997, p. 601, see the letter of Bonnand to Bissachere written from Pondicherry dated 11 February 1825. The text runs: 'jai beaucoup de la piene a prononcer ces letters toute baroques'.
92. AMEP, Correspondance Pondichéry-Paris, Lettres, vol. 997, p. 772, see the letter of Louis Charles Hebert written from Pondicherry dated June 1826 to Langlois in Paris.
93. S. Jeyaseela Stephen, *Louis Savinien Dupuis (1806-1874): A Biography of the MEP Missionary in Pondicherry*, Pondicherry, 2017, pp. 24-5.
94. AMEP, Correspondance Pondichéry-Paris, Lettres, vol. 999, see the letter of L.S. Dupuis dated 18 November 1839 written from Ballapuram to the MEP Superior in Paris.
95. AMEP, Correspondance Pondichéry-Paris, Lettres, vol. 999, see the letter of L.M. Mousset dated 16 September 1835 written from Pondicherry to Jean Tesson in Paris.
96. AMEP, Correspondance Pondichéry-Paris, Lettres, vol. 999, see the letter of L.M. Mousset dated 29 November 1836 from Sarugani to Jean Tesson in Paris.
97. AMEP, Correspondance Pondichéry-Paris, Lettres, vol. 999, see the letter Bonnand the Bishop of Pondicherry dated 18 February 1841 to Langlois in Paris.
98. Société Archeologique de Sens (hereafter SAS), *Dictionnaire Tamoul Français par deux missionaries Apostoliques de la Congregation des Missions-Etrangères*, Pondichéry, 1855, see the preface, pp. viii-ix.
99. Ibid.

100. Ibid., pp. ix-x.
101. Ibid.
102. Ibid., p. xii.
103. Ibid.

Part - I

From Pre-Modern to Modern Tamil Language and Grammar in Missionary Dialogue

3

Studying Tamil Grammar, Generating Grammars in European Languages and Circulating Paper Manuscripts 1549-1855

The European missionaries after coming to know about the Tamil literature they understood that a grammar of that language exists. They found difficulty in learning Tamil at the beginning in the absence of learning tools. They clearly understood that *ilakkanam* (classical grammar) was the norm for *ilakkiyam* (classical literature). They further learnt the earliest Tamil grammar work was called Tholkaapiyam wherein Tamil grammarians and poets have codified strict rules for poetic conventions, besides for the linguistic and literary usages. Europeans could not understand Tamil grammar at all since it was found only in poetry. They realized then that they were free to choose the Latin and Greek grammar patterns while learning Tamil. They began to compare the literary forms whatever they met with in Tamil and composed Tamil grammar in Portuguese, Dutch, French and Latin.

Portuguese merchants and missionaries had first come into contact with the parava fishermen who were considered low in the highly caste ridden Tamil society. The missionaries came to know about the highly colloquial form of Tamil which was spoken by these paravas. But there was a highly literary form of Tamil also in use among the upper castes inherited as the legacy of the Sangam period literature. The first conversion of the Tamils on the coast began in 1536 and the real missionary efforts of learning Tamil had started immediately after that. At that hour the missionaries confronted great difficulty in communicating with the locals. It was felt necessary by them to master the spoken language, at first in order to convey the message of Catholic Christianity. They desired to preach the tenets of Catholicism in spoken Tamil. Their next task was to improve their knowledge in literary

Tamil. They formulated a policy and devoted their attention to learn Tamil grammar properly and systematically.

It must be admitted that Portuguese missionaries did not have the skill over poetry and so assigned top priority to colloquial spoken Tamil. Instead of learning Tamil grammar through proper methods by using grammatical treatises, they adopted their own self learning which became the need of the hour from social and intellectual point of view. Portuguese missionaries developed interest to learn Tamil grammar to obtain skill over the spoken Tamil, rather than written literature in the beginning. Thus the earliest phase and the process of modernization of the Tamil language and grammar were initiated by the Portuguese missionaries. The Euro-Tamil projects started on the subject of Tamil grammar are available only in manuscript form scattered and spread in various archives in Europe and India. Hence the present chapter attempts to explore the contribution of the Europeans in the field of Tamil grammar.

Tamil Grammar in Portuguese

Paulo della Valle the Italian Jesuit who sailed from Lisbon to Goa had narrated his voyage and the difficulties experienced.[1] He was posted at Vembar and worked along with Xavier and Francisco Gonçalves. Paulo della Valle had written a grammar of Tamil. Anrique Anriquez had mentioned it in his letter dated 21 November 1549 that he began to use it for learning Tamil.[2] We have no trace of his manuscript except the reference that he died of exhaustion in Punnaikayal on 4 March 1552.[3]

Anrique Anriquez who began to learn Tamil decided to compose a Tamil grammar. He stated that he was requested by Xavier in 1548 to write a grammar work based on the model of the Latin and Greek grammars.[4] It is noted from the letters of Anriquez subsequently written that he had taken the advice of Xavier very seriously and he was busily engaged in learning Tamil grammar. In 1549 Anriquez wrote thus: 'When I came to this coast I wanted to learn this language so that I could read it. I determined to learn the language. And so, day and night, though not giving up my visitation of the places in my charge, I did nothing else, and God was pleased to help me. To learn it, I used a grammatical pattern they use in Latin to learn the conjugations. In this way I worked to learn this language. I conjugated the verbs. To find, the past-participles, futures, infinitives, subjunctives, etc., cost me much labour; also to learn the accusatives, genitives, datives and the other cases. I also had to learn which (subject, predicate and object) to put first'.[5] Thus to prepare this Tamil grammar, Anriquez had drawn the basics from the language spoken by the paravas. He structured the spoken

Tamil on the grammatical pattern of Latin. When he began to learn Tamil, he wrote it down in Portuguese phonetics.

Anriquez in his letter dated 13 November 1549 to Ignatius of Loyola mentioned that his Tamil grammar work had been completed.[6] He had enclosed a rough draft of the grammar along with the letter and awaited decision to know whether or not it could be printed. He expressed his desire to bring out the same in print within two years by which time he could make a perfect grammar with some more additions to it.[7] He also wanted to omit some parts in the manuscript copy of the Tamil grammar he had submitted to Portugal. He however preferred to send another draft that would be more legible, neat, and complete. He himself thought that the manuscript he had submitted was unsuitable for publication as it stood.[8]

Anriquez mentioned in his letter from Cochin dated 12 January 1551 to Simão Rodrigues in Portugal that if he wanted to learn the Tamil language particularly to read and write he can do it easily with the grammar that he had prepared. He also said that with the established exercises given in the manual he could learn to speak Tamil.[9] Fr João de Mesquita in his letter in 1558 wrote that Anrique Anriquez had composed a grammar (arte) in Tamil language which had been a great help for those who wished to learn the language He stated that Anriquez spoke well and he no longer needed interpreters.[10] Anriquez himself had written about his own actions in 1558. He stated that some years back he sent a copy of the Tamil grammar, and he had improved on it by now, but could not finish it due to his many occupations.[11] Over a period of 17 years we find that Anriquez made continuous efforts to improve upon the grammar that he was doing as evident from his letter dated 24 December 1564 written from Punnaikayal.[12] He had by this time reached the stage when he could state with confidence in his letter written from Punnaikayal dated 31 December 1567 as follows: 'I have the feeling that I now have the ability to compose a grammar in any language if I could have an interpreter or two and a scribe in that language'.[13] Further he mentions that for his Tamil grammar to be understood more easily, it is necessary to have a knowledge of Latin grammar. Those who do not know Latin should read through with the help of the Portuguese grammar of João de Barros.[14] Anriquez has stated that after working many years on the Tamil grammar, he could achieve some measure of perfection only in the year 1566. He must have thus worked doggedly to accomplish this formidable task.[15]

The medieval Tamil grammar Nannul written in verse during the thirteenth century was available to Anriquez. He tells us that he devised his own method and mastered it by spending sleepless nights. He had thus succeeded in simplifying Tamil grammar for the Portuguese to learn the language. He revised his Tamil grammar work again in 1565. By this time, he improved his vocabulary to more than 2000 words. He said in one of his letters that his command over Latin facilitated him to write Tamil grammar within a given time.[16]

Anriquez in his letter dated 31 December 1567 mentioned that he had written a major part of the Tamil grammar although it had not been finished, because he lacked scribes to perfect it. He said that a deacon who was learning Tamil language saw some of the things that he had written and affirmed that it was better than the grammar that was written in Konkani in Goa, and there they also collected some rules for the grammar. This brother was one of those who were learning Konkani at that time. The grammar was composed in Goa by one or two native Konkanis who knew Latin. Anriquez hoped that this Konkani grammar under preparation in Goa would be improved to the level of the Tamil one, on which he had worked for many years. H stated that few words have been similar in both languages, but the structure of the sentences was very similar. Anriquez hoped to send this Tamil grammar to Goa soon in order to have it examined. The truth at that point was that he was not able to improve the Tamil grammar for he had no adequate time and the help of the scribes.[17]

A Tamil grammar manuscript preserved at the Biblioteca Nacional de Lisboã, Portugal, is attributed to Anriquez since he had sent it to Portugal. Based on the internal and external evidences it is also attributed to Anriquez.[18] This manuscript titled 'A Arte de Lingua Malavar' has a total of 159 + 1 folios of 213 x 153 cm size. The Tamil examples are all given in Roman transcription. Anriquez might have made additions, deletions and rearranged the text. The folios therefore have been numbered and then renumbered as we notice that folio 146 also bears the new folio number 156, though according to the old folio number, the manuscript ended at fl.150. This grammar however never got printed. Anriquez is the second European to compose grammar of Tamil in prose in detail. All the missionaries who followed to come and preach the gospel in Tamil country therefore used his Tamil grammar.

Anriquez claimed that he could construct sentences in Tamil in a very simple manner better than the Tamils. Further he stated that he worked on the Tamil grammar so that the priests may learn it easily. He worked on the conjugations, declensions and the rules of grammar, so that one could learn Tamil without a teacher and even independently. From the Spanish preface that he wrote to the book Flos Sanctorum printed in 1586 we understand that he had already completed writing the Tamil grammar.

It is gleaned from the manuscript preserved in the Biblioteca Nacional de Lisboã in Portugal from folio 7v that this Tamil grammar is based on the model of Latin grammar. The manuscript has folios written in recto and verso and it is in Portuguese with occasional digressions into Latin. All the pages begin with *Sesu Maria* (Jesus and Mary). Interestingly in the section on nouns we find the following inscription: 'In the name of our Lord Jesus Christ'. The illustration of Tamil words is found in the Roman alphabet for which the rendering in Tamil script is frequently given. The style of the grammar is personal as spoken directly to a student. The Tamil grammar begins with a detailed table of contents.[19] There is a short introduction where the author explains the Tamil alphabet about how it is to be pronounced.[20] The long and short vowels are distinguished in this grammar book.[21]

The section on nouns begins with the caption 'In the name of Our Lord Jesus Christ'. The treatment of the noun forms are found in the first section of the grammar. The author had devised six cases such as, nominative, vocative, accusative, genitive, dative and ablative. He also proceeds to develop eleven tenses. The pronouns are given (fls.30-31). Then we find the declensions (fls.23v-25v). The examples cited are given as below. He used the term pandam and gave the declensions.

Singular

Nominative - Pandam of facto	- Pandam
Genitive - Pandam vrhaeia, do facto	- Pandamudaya
Dative - Pandatucu, ao facto ou para o facto	- Pandathuthukku
Accusative - Pandataie e tabe pandam, of cato	- Pandathai, pandam Vocative
Pandame, o facto	- Pandame
Ablative - Pandatile etabe pandatil, e o facto	- Pandathiley

Plural

Nominative - Pandangal, os factos	- Pandangal
Dative - Pandangalucu, aos fatos ou para os fatos ou para os fatos	- Pandangalukku
Accusative - Pandangallaei, os factos	- Pandangalai
Ablative- Pandangalile, e os fatos	- Pandangaliley

SOURCE: Bibliotheca Nacional de Lisboã, Portugal, MS 'A Arte de Lingua Malavar', Codice 3141, fl.17r; See also Bibliothèque Nationale, Rue Richelieu, Paris, MSS Indien, no. 188, fls.5-10.

In order to make the reader understand better Anriquez gave another example. He began to use another term 'raasaa' and made the declensions as follows.

Singular

Nominative - Jraia	- Raasaa
Genitive - Jraia vrhaeia, del Rei	- Raasaa udaya
Dative - Jraiaucu, ao Rei oupara el Rei	- Raasaavukku
Ablative - Jraiauaei, O rei	- Raasaaivai
Vocative- Jraiaue, O Rei	- Raasaavae

Plural

Nominative - Jraiacal, os Reis	- Rassaakkal
Dative - Jraiacalucu, aos Reis ou para os Reis	- Raasaakkalulukku
Accusative - Jraiacalaei, os Reis	- Raasaakkalai
Vocative - Jraiacale, os Reis	- Raasaakkaley

SOURCE: Biblioteca Nacional de Lisboã, MS 'A Arte de Lingua Malavar', Codice 3141, fl.18r; Bibliothèque Nationale, Paris, MSS Indien, no.188, fls.5-6 and 8.

The participles are discussed at length and Henriques uses the term visuvathi. The present participle is given as follows.

Nominative - Vichuuadiquiraen, o que cree ou o que crer	- vichchuvathikiravan
Gentitive –Vichuuadiquirauaen, vrhaeia, do que cree	- vichchuvathikiravanaudaya
Dative - Vichuuadiquirauaencu, ao que cree ou para o que cree	- vichchuvathikiravanukku
Accusative - Vichuuadiquirauaenaei, o que cree	- vichchuvathikiravanai
Vocative - Vichuuadiquirauene, o o que cree	- vichchuvathikiravanae

Then he proceeds to give for the masculine gender, singular, feminine gender, singular, and respectful forms singular, and then the plural forms of these and the future negative.

The conjugations (fls.41-42) and infinitives (fls.33-34) are also found in the grammar book. Anriquez gives examples such as '*Vai dizer a Pedro q venha logoca*' (Go tell Peter to come soon) and in Tamil '*Pedro quitta poy avoenoei chardinu inge vara cholu*'. The grammar book very abruptly ends and it is possible that the work was in progress and it was not completed at that time.[22]

In the Portuguese-Tamil grammar of Anriquez the pronunciation of Tamil words is given in the Roman alphabet to which a rendering in Tamil script is frequently added. The grammar begins with a fairly well detailed 'Table of Contents' on folios two and three. The page numbers given in the 'Contents', do not tally with the actual pagination of the text. This discrepancy could perhaps be attributed to the fact that the grammar may have been in the process of completion. In his introductory notes on folios five and six, he explains the pronunciation of Tamil vowels and consonants.

The first part of the grammar of Anriquez deals with Tamil words and phrases which are given in Latin accompanied by their corresponding Tamil spelling. Although the difference between the short and long vowels in Tamil, contrary to Portuguese pronunciation, is mentioned in the sixth folio, the author had however failed to mark the phonological difference in the text. Further the representation of r (ra) and R (ra) is also arbitrary in the manuscript. Adjectives are explained in folios 26r-27v. The conjugation of verbs has been dealt with elaborately in folios 37r to 142v. The author has given all the five declensions of the nouns with their variations under the nominative, vocative, accusative, genitive, dative and ablative cases. Syntax is explained in the last nine folios. The Tamil words used in this grammar amounts to 310 in number.[23] In the manuscript we find that no dots (*pulli*) are used over consonants. The manuscript is incomplete and abruptly ends with folio.159v. In the grammar he uses only commonly spoken Tamil words, since the purpose was to teach the missionaries to speak the Tamil language fluently rather than using it for writing.

The manuscript of Portuguese-Tamil grammar of Anriquez has been the subject of study and many have pointed out the merits and defects of this grammar. Anriquez had put the pronouns under the category of nouns. Although on folio 5v, he mentioned that he would distinguish between the long and short vowels by putting some mark over the long vowel, the author had rarely done so. Anriquez had just mentioned the verbal compounds but does not explain the difference between them and the simple verbs in folio 48r and even suggests it as a gerund in folios 48r, 141r and 140v.[24]

Anriquez deals with the nouns in the first section of his grammar. The author has identified the five declensions of the noun ending with an, al, am, tay, tu, and ru in relation to the six cases, viz. nominative, vocative, accusative, genitive, dative and ablative. He explains the rules for the declension of nouns in folios 23v-25v. He points out (in folio 25r) that the Tamils are not accustomed to using the plural form unlike the Portuguese. Adjectives are dealt with in folios 26r-27v. The author is aware of two categories of verbs which he calls variously as personal and impersonal, rational and irrational, regular and irregular. He gives eleven different variations of the tenses starting from the present tense along with their meanings. He also gives a summary of the rules for the use of cases. He makes a passing reference to the degrees of comparison and how they are formed in Tamil. Thereafter the manuscript ends in folio.159v. Towards the close, Anriquez states that he would like to add a few more things in the text. He wrote Tamil grammar in Portuguese and had based on two grammars, one Latin and one Portuguese, by João de Barros.[25] It is clear that he began his work of Tamil grammar in 1548.[26] For almost 16 years, in many of his letters sent to Europe, he reported on the improvements made from time to time and finally in 1564 he wrote to Diogo Laínez, the superior that he had accomplished his task.[27] The Tamil grammar manuscript of Anriquez is also available in the State central Library of Goa and it is found along with the Portuguese-Tamil dictionary compiled by him.[28]

Gaspar da Aguilar (1588-1648) joined the Society of Jesus in 1609 and he reached Cochin in January 1610. He went to Colombo and Jaffna in 1620. He was at Santhome of Mylapore in 1623, at Cochin in 1627 and in 1634 he was at Punnaikayal. He taught grammar and theology in the seminary in Cochin. He was dismissed from the Society of Jesus in 1645 for his misbehaviour.[29]

From a record we find that Gaspar da Aguilar spoke Tamil very fluently and so he was sent from Tuticorin to Madurai to hold discussions with Tirumalai Nayak the ruler on 22 March 1638 regarding the purchase of saltpetre by the Portuguese.[30] It is reported later by Balthasar da Costa another Portuguese Jesuit that Gaspar da Aguilar had written a more advanced grammar in Tamil which he found more helpful to learn Tamil.[31] The manuscript grammar of Gaspar da Aguilar is now preserved at the Hamburg University Library.[32] Philippus Baldaeus a Dutch missionary who had worked in Nagapattinam and Tuticorin in 1660-61 also mentions about Gaspar da Aguilar, and his Portuguese-Tamil grammar which he used.[33] According to Jesuit correspondences when Aguilar was in Punnaikayal he had completed the work around 1634. Baldaeus who wrote a short Tamil grammar in the Dutch language got it printed in 1672 (being the first Tamil grammar to be printed in any European language) mentions in his introduction that the Tamil grammar work of Gaspar da Aguilar was good.[34]

Jean-Venant Bouchet (1655-1732), a Jesuit in his letter written from Pondicherry to Père Du Halde dated 18 December 1714 mentioned that a Tamil grammar manuscript was being sent to the library of the king of

France.[35] We have therefore reasons to believe that the manuscript sent to France could be that of Gaspar da Aguilar now available among the collection of manuscripts at the Bibliotheque Nationale in Paris.[36] The name of the author and the date is not found mentioned but the text of the manuscript records that the French who were initially in Masulipatnam (in 1672) attempted to learn the Tamil language with the help of this manuscript.

This copy of the Portuguese-Tamil grammar of Gaspar da Aguilar is written on both sides of the paper. The sheets counted are fifty-seven. The manuscript begins with Tamil vowels and consonants on folio one and ends on folio 3r. The Sanskrit alphabets commonly used in Tamil are given on the fourth folio. The declension of nouns start on folio five and all the variations of Tamil vowels and consonants are given on folio seven along with the numerals from one to ten besides 100 and 1000. The author has not used dots over consonants as was the practice at that time. The first declension starts on folio eight with the word raasaa. For the second declension the word toppi is used and its plural form is also given. The word kambu is used for the third declension on folio ten. The word pandam is used for the fourth declension. Further, adjectives are explained with the use of the word nallavan and we find here both the singular and plural forms. The comparative and the superlative degrees are explained on folio 16v. The number names from one to ten find place on folio seventeen. The verbs are explained on folio twenty-four. While the participles find place on folio 45v the infinitives are found on folio fifty-two. The prepositions are given on folio 53v. The manuscript finally ends on folio fifty seven. In all, folios 44v and 45 alone are left blank. Since the manuscript is not the same as that of Anrique Anriquez it may possibly be concluded that this Paris manuscript is certainly that of Gaspar da Aguilar and it was widely used by the missionaries of that time and might have been sent from Pondicherry to France. Gasper da Aguilar's grammar had been cited by Philippus Baldaeus and also Balthasar da Costa for they had used.

Balthasar da Costa learnt Tamil very well and he had an excellent command over the language.[37] He composed a Tamil grammar (Arte Tamulica) and completed it in 1656. A copy of the manuscript prepared by someone and completed in 1731 is preserved in the State Central Library in Goa.[38] This grammar copy has 49 pages. The title page and the first page is found missing. The manuscript is attached to the Portuguese-Tamil dictionary compiled by him and therefore must have escaped the attention of scholars for a long time. Balthasar da Costa in his Tamil grammar mentions that his work was mainly based on the earlier grammar of Gaspar da Aguilar. He therefore instructs the learners needing more detailed information to refer to the grammar of Gaspar da Aguilar.

Another copy of the Tamil grammar manuscript of Balthasar da Costa is available in the British Library, London.[39] It is in good condition and we find in the beginning Da Costa explains how Portuguese speakers would have to pronounce all alphabets. Then the proper section on Tamil grammar consists of six chapters. The first chapter deals with various forms of Tamil nouns and pronouns, adjectives and genders. The second chapter the longest chapter of all and it illustrates conjugation of the verb visuvasiththal (to believe) in its affirmative, honorific, negative singular and plural forms, various tenses, imperatives and moods. The third chapter is on the passive voice and it is very brief. The fourth chapter illustrates various ways to compose verbs. The fifth chapter shows how some substantives and their regular plurals are formed. The sixth chapter explains how the Tamil language does not have any prepositions but only postpositions. An appendix includes examples of certain future verb forms imperatives and infinitives.

The first chapter deals with nouns, names of places, and gender. He uses the term karthan-karthar, jeyam-jeyangal, veedu-veedugal and aaru-aarugal and he provides the various cases for these words. He also uses the words naan-naangal-naam; nee-neengal-neer; avazhlavarghal-avar; athu-athugal; thaan-thaam-thaangal and gave the various forms of the cases. He mentions the several forms of words such as avar-idathil, avar-aalaey, avar-odey, avaraipaarthu, avaraikurithu, avarukkaaha, avaraikondu and points out that Brahmins alone used the terms avaal and avaaludaya.

He states that there is no adjective (*peyar yecham*) in Tamil as found in the European languages. These are common for all genders. These will appear before the names. The conjugation of the verbs had been dealt with the use of the word visvasi. The details of singular and plural mentioned are as follows.

Vichuvasiquiren - I believe
Vichuvasiquirai - You believe
Vichuvasiquirir - You believe (respect)
Vichuvasiquiran - He believes
Vichuvasiquirar - He believes (respect)
Vichuvasiquirom - We believe
Vichuvasiquirigal - You believe
Vichuvasiqurargal - They believe

It appears that Balthasar da Costa had elaborated upon these forms mainly taken from the grammar of Anrique Anriquez for the same word vichuvasiquiren is found.[40] It appears that details of the forms of the same word are also taken from Anriquez. Thus it is possible that he was aware of the grammar work of Anriquez.

In the last page of Balthasar da Costa's grammar manuscript preserved in Goa, there is a note of a later date declaring that this particular copy belonged to one Fr António das Chegas, OESA, an Augustinian monk. It appears that Balthasar da Costa's manuscript of Tamil grammar was used by Fr António das Chegas who was ordained a priest on 27 December 1787 to learn Tamil. He was at the house of St. Francis in 1797.[41] He took an oath of allegiance to serve in the missionary areas belonging to the British on 11 April 1803.[42] In 1806 he was the parish-priest at Manapadu and in 1807 at Periya Thazhai.[43] In Nagapattinam he was instructing the Portuguese in Tamil on 20 November 1802.[44] He was at Mylapore in October 1810.[45] Thus the copy of Balthazar da Costa's Tamil grammar now preserved in Goa Library was used by António da Chegas. Thus we come to understand that the Tamil grammar of Balthasar da Costa was very popular among the fellow missionaries even after a century.

Balthasar da Costa discussed the relative pronouns and compared the same with Portuguese and Latin.[46] He maintains that nouns common to all genders always occur before the names. Then he proceeds to deal with the masculine, feminine and neuter genders and explains here the usage of the expression Devarir sollitru and points out that the usage occurs only in masculine singular. Chapter two deals with conjugation of verbs beginning with the present tense, singular and plural and uses the example vichuvathikiran. He then goes on to explain the negative forms and later the past tense of the same verb. Chapter three deals with the active voice and the passive voice and the word muhkikiran (modern usage mudikiran) is used as an example. Chapter four deals with joint verbs (kuttu vinai), command verbs (eval vinai), and also the combination of both. Chapter five of his grammar deals with auxiliary verbs and irregular verbs. In chapter six, Balthasar Da Costa deals with the prepositions, adverbs, interjections, conjunctions and disjunctions.

Some of the salient features of Balthasar da Costa's grammar may be worth mentioning here. He explains how a finite verb is changed into a noun in Tamil, such as visuvasikkrathu, visuvasithathu, visuvasippathu, and visuvasiyathathu. Here he used the present participle used by Anrique Anriquez (vide folios 25 to 30) using the same word visuvasi.

Balthasar da Costa also gives a list of expressions in Tamil which were considered amangalam (inauspicious) and we may give an example here such as muhikiran instead of mudikiran. He then refers to irregular verbs and distinguishes the root difference, in addition to finite verbs and verbal nouns. He while explaining the prepositions in chapter six, records that the grammar of Gasper da Aguilar, a predecessor, should be consulted for further details.

Antão Proença the Portuguese missionary wrote Tamil grammar in Portuguese in 1665. The details are found in a manuscript copy made in 1670 now preserved at the State Central Library in Goa.[47] Proença had given five chapters. The first chapter deals with formation and structure of the Tamil alphabets and pronunciation. In Chapter two we find the details of Tamil grammar composed by Balthasar da Costa. In chapter three he explains the past tense, future tense, command, and adverbs with rules. In chapter four he gives the details of magara yeeru and udampady mey. In chapter five he outlines those Portuguese alphabets that would come under the related Tamil alphabets. It must be said that those who printed the dictionary of Antão Proença in 1679 had omitted his Tamil grammar found in Portuguese as it was attached to the manuscript submitted for approval and accorded sanction for printing the dictionary through permission by the inquisition in Goa. Constanzo Giuseppe Beschi, the Jesuit was aware of the grammar work of Antão Proença. He in his grammar in Latin for spoken Tamil written in 1728 and got printed at Tranquebar press in 1738 (vide paragraph 12) notes that the term kannu and kanru used by brahmins and other castes which Antão Proença had cited in his grammar was incorrect. A copy of the Tamil grammar of Antão Proença was made and sent to Paris, is preserved in the National Library, Paris.[48]

Portuguese missionaries who set themselves to learning Tamil realized soon that Tamil grammar in verse written by the Tamil poets would be difficult to comprehend. Hence based on Greek and Latin they wrote a comparative grammar of Tamil to facilitate in a larger measure to master Tamil quickly and systematically. Hence they began to write down the elements of Tamil grammar in Portuguese in their own mother tongue. Printing books in Tamil that dealt with the Christian tenets had been given priority so as to reach the people. As only limited number of copies of Portuguese-Tamil grammar was required, and that too for the use of the missionaries alone, no attempt was made by them to print the same. It is very important to note how each of these Portuguese missionaries evolved their own method of learning and authoring Tamil grammar by understanding the rich grammatical tradition of Europe and not Tamil country.

Tamil Grammar in Dutch

Tamil grammar came to be written in Dutch language and the age reached a new epoch of history. Since Tamil grammar was written in Portuguese language, the Dutch used the same old techniques for writing and we find they copied or adopted or followed the same model without any change and used the model of Greek and Latin grammatical models that suited to Tamil grammatical tradition.

Philippus Baldaeus the Dutch missionary who worked in Tamil country and Sri Lanka wrote introduction to Tamil language and it is dated 21 July 1671. He wrote it from Geervliet in Holland after his return.[49] We find his Reebe unleitung da der Malabrsichen Sprachtonft and Malabarische Sprachfort folgen einige Borefellengen as der Malabarcihen sprachtum. He mentions cases such as Pradaemi (nominative), Datier (gentiven) Trieti (dative) accusative (chadurti) Parcham (vocative) and Xasti (ablative). Baldaeus uses the word vannaan in his first declension and conjugation.[50] We find also prayers such as Pater Noster, and the Creed at the end. The prayer Pater Noster (vaaanagaliley irukira) has no dot on the consonants. There is an appendix and has three tables. Korte malabsrsche letter-konst (short Tamil orthography) is given. There is no difference we find between the long and short o, ye, ko and key. In 1672 his grammar was translated into German and got published. In 1704 it was translated again into English and also got printed.

Baldaeus mentions some rules for the attaining the knowledge of the Tamil language. The first declension among the Tamils comprehends the nouns only of the masculine gender, which terminate in the singular number in n, and in the plural in r. He mentioned that Tamils have their singular and plural numbers like the Dutch and the named their case as follows:

The nominative - Pradamei
Genitive - Dutiei
Dative - Tritei
Accusative - Chadurti
Vocative - Panchami
Ablative - Xasti

Baldaeus gave an example of the first declension, singular through the use of the word vannaan.

Nominative - vannaan, Fullo, The fuller (of cloth)
Genitive - vannaaureja, the fuller
Dative - vannaanucu, to the fuller

Accusative	-	vannaaner, the fuller
Vocative	-	vannaano, o you fuller,
Ablative	-	vannaaniratil, of the fuller

He also said that it is to be observed that Tamils have four sorts of ablative cases: (1) the locais, as the last mentioned. (2) Causalis, vanaanale, because of the fuller. (3) Socialis, vanaanore, with the fuller. (4) Comparationis, vanaanil, in comparison with the fuller.

Baldeaus also gave the plural of through the use of the word Vannar

Nominative	-	vannaar, fullerone, the fullers
Genitive	-	vannanareja, the fullers
Dative	-	vannaarcul, vanaaruccu, to the fuller
Accusative	-	vannaarei, the fullers
Vocative	-	vannaare, o you fullers

Ablative (1) vannaar iratil, vannaar iratile, of the fullers; (2) Vanbaaral, vannaarale, because of the fullers; (3) Vannaarore, vannaarorum, with the fullers; (4) Vannaaril, vannaarilum, in comparison of the fullers.

Baldaeus gave an example of conjugation in the affirmative. This portion is largely taken from the grammar work of Balthasar da Costa and the word used is visuvasikiren.[51]

The present singular tense

Vichuvadigirren	-	I believe
Vichuvadiguirray	-	You believe
Vichuvadiguirran	-	He believes
Vichivadiguirral	-	She believes
Vichuvadiguirardu or vichuvadiguidu	-	It believes

The plural

Vichuvadiguirrom	-	We believe
Vichuvadirirgal	-	You believe
Vichuvadiguirargal masculine/ feminine	-	You (men and women) believe
Vichuvadicradu or vichuvadiguidu	-	the things believe

In the negative: The present singular tense

Vichuvadichren illei	-	I do not believe it
Vichuvadicrei illei	-	You do not believe it
Vichuvadiezan illei	-	They do not believe it
Vichuvadicrad illei	-	It does not believe it

The plural

Vichuvadicrem illei	- We do not believe it
Vichuvadicrirgal illei	- You do not believe it
Vichuvadicrargal illei	- They (masculine and feminine) do not believe it
Vichuvadicrad illei	- They (neutral)
Vichuvadiguid illei	- Do not believe it

Baldaeus gave the present tense of honour, i.e., when you speak to persons of quality.

In the affirmative

Vichuvadicrora	- We (our majesty highness) do not believe it
Vichuvadicrir	- You (your majesty) do not believe it
Vichuvadicrare	- They (their majesty) do not believe it

In the negative

Vichuvidcron illei	- We (our majesty) do not believe it
Vichuvadicrir illei	- We (our majesty) do not believe it
Vichuvadicrar illei	- They (our majesty) do not believe it.

The imperfect tense

Vichuvadicrane apo vichuvadieren	- I believed
Vichuvadicreye aop vichuvadicrom	- We did believe it.

The perfect tense

Vichuvaditen	- I have believed
Vichuvaditom	- We have believed

The past prefect tense

Aducu mune vichuvaditay	- I have believed
Aducu mune vichuvaditon	- We had believed

The future tense

Vichuvadipen	- I will believe
Vichuvadipon	- We will believe

SOURCE: Philippus Baldaeus, *Nauwkeurige beschryvinge van Malabar en Choromandel, derhalver aangrenzende ryken, en het machtige eyland Ceylon. Nevens een omstandige en grondigh doorzochte ontdekking en wederlegginge van de afgoderye der Oost-Indische heydenen*, 3 vols, Amsterdam, 1672, pp.195-7; See also, *A True and Exact Description of the Most Celebrated East India Coasts of Malabar and Coromandel as well as of the Isle of Ceylon with their Adjacent Kingdom and Provinces*, London,1703, pp. 663-65.

It may be mentioned that Baldaeus had used at length the details furnished by Anrique Anriquez in his grammar (vide folios. 25 to 30) by using the same word Vichuvadigirren and also used the present participle of Anriquez by using the same word.

Baldaeus mentioned that whence it is evident that the Tamil language is very difficult to be learnt, especially by reason of the vast number of words where of it consists, one and the same thing being frequently expressed by diverse words and names. Thus not only every day in the week, but also throughout the year, has its peculiar name. It is further manifests from hence, that the Indians are not very unpolished as some Europeans represent them, and that they treat one another (especially persons of quality) with singular civility and respect. It may be said that Baldaeus work is not a complete work on grammar. He gave a brief sketch when he wrote about Tamil country he mentioned in a chapter on aspects of Tamil grammar.[52]

Tamil Grammar in German

German Protestant missionaries who came to preach the Gospel in Tranquebar at first learnt Tamil language with the help of *dubashis* (interpreters) and *vaathiyars* (teachers). They carried on their apostolic work with the help of *upadesiyars* (catechists). Their task was to teach the spoken Tamil language to the fellow missionaries in Europe who had opted to work among the natives, who would find it difficult to be conversant with the literary Tamil forms of grammar. In fact there was no prose form of writing grammar in Tamil. When such a daunting task came to teach the language to their fellow Protestant missionaries, the idea of writing Tamil grammar in German was done by Bartholomäus Ziegenbalg.

On 22 September 1707 Johann Sigismund Hassius the governor of Tranquebar had procured a new copy of a Tamil grammar in Portuguese written by a Catholic missionary. It was presented to Ziegenbalg. Mudaliappa taught Ziegenbalg and Pluetschau from this Tamil grammar in Portuguese composed by the Catholic missionary.[53] Ziegenbalg acknowledged that he used it and as on 25 September 1706 he was able to improve his learning Tamil and he spoke Tamil by conducting catechism classes.[54] Ziegenbalg began to write Tamil grammar in German. He derived basic grammatical principles directly from reading Tamil texts. He put them together and translated them into German in 1713. He used the two hymns that the Tamil Lutherans sung at that time to explain his grammar rules. He sent his manuscript to Frederick IV the king of Denmark and also dedicated this work to the king.[55]

Another copy of this Tamil grammar in German was prepared and kept in Tranquebar which had existed until 1730. Christoph Theodosius Walther who prepared a catalogue of all Tamil books in 1731 remarked that Ziegenbalg had compiled this grammar entitled Einleitung zur Erlernung der Malabrischen Sprache (Introduction to learning the Tamil language). This manuscript contained three chapters such as (a) Kurza und grundliche Grammatikalische instruction (short but through grammatical instructions), (b) Eine unter den Malabaren gehaltene Oration (a sermon held among the Tamils) and (c) Ein Vokabularium zu dieser Oration (a catalogue of words used in this sermon).[56]

Ziegenbalg's grammar manuscript sent to the king of Denmark in deed helped the new missionaries who sailed from Copenhagen. According to records they brought a copy of it with them and used during the voyage and also after landing in Tranquebar to learn Tamil.[57]

The Grammar for Literary Tamil in Latin by Costanzo Giuseppe Beschi, 1730

Beschi wrote a grammar titled in Latin Grammatica Latino Tamulica ubi de elegantiori linguae Tamulicae (titled in Tamil Senthamizh) that dealt with literary Tamil or higher Tamil which was completed on 13 September 1730. He wrote this grammar of High-Tamil which was more or less an addition to his grammar of the lesser dialect. Beschi in his grammar of High-Tamil gave an introduction to the mysteries of classical literature and stated 'ex indigenis ipsis vix aliquot jam invenies qui elegantiorem hanc linguam perfecte calleant, qui prima ejus rudimenta alligerit, suspicantur caeteri; et siquid ex reconditis eorum voluminibus in medium proferas, pendent ab ore omnes. Si vero quae ipsi vix a longe attigere potuerunt, exterum hominem exhausisse viderint, quibus laudibus non extollent? Facile Magistrum audient quem doctissimum admirantur.[58]

Beschi had extensively cited Thirukural in the grammar of Senthamizh and the details are as follows:

In Latin Original					*Kural number*	
Page	1	para	1		1	Agara muthala vezhuththellaa
Page	1	para	1		80	Abin vazhiya thuyir nilai
Page	1	para	5		66	Kulalinithi yaazhini thenba
Page	3	para	6		2	Katrrathanaalaaya payenenkol
Page	4	para	8		664	Solluthal yaarku
Page	6	para	14		151	Agazhvaarai thaang nilam
Page	13	para	21		964	Thaliyin nilanztha mairanayar

Page 16	para 23	408	Nallaar katpatta varumai
Page 23	para 33	26	Seyarkariya seyvaar
Page 26	para 37	450	Pallaar pagai kozhlarir
Page 27	para 37	513	Anbarivu thettra mavaavinmai
Page 35	para 50	57	Sirai kaakum kaapeven
Page 35	para 50	763	Olithakkaa lenna muvari
Page 44	para 57	660	Salthaar porul seythe
Page 47	para 62	439	Viyavarka vengnaanru
Page 54	para 69	208	Theeyavai seythaar kedutha
Page 55	para 70	90	Moppa kulaiyu manichcha
Page 58	para 72	26	Seyarkariya seyvaar
Page 59	para 73	366	Anjuva thoru marane
Page 62	para 76	426	Evvava thuraiva thulaga
Page 69	para 89	114	Thakkaar thagavila
Page 71	para 95	393	Kannudaiya renbavar
Page 72	para 95	20	Neerintr amaiyaa thulagenin
Page 72	para 95	34	Manathukkan maasila
Page 72	para 95	453	Manathaana maandar kunarchi
Page 75	para 101	667	Uruvu kandellaamai vendu
Page 75	para 101	32	Arathinung kaakamumillai

It may be worthwhile to study the formation and structure of language and the differences pointed out between Senthamizh and Kodunthamizh by Beschi in a comparative framework in the light of his two works.

Subject	***Place***	***Senthamizh***	***Kodunthamizh***
Yezhuthu	1.15	-	Grantha letter
Yezhuthu	1.15	Aytham	-
Yezhuthu	2.17	-	Letters that would come before ra, na, la, ya
Yezhuthu	2.19	Uiralapadai	-
		Ottralapadai	-
Sixth case	9.34	Athu	Udaiya
Seventh case	9.35	Kann	Ill
Present tense	26.67	Illai	Undu
Viguthi thoguthal		Vanagukira	Vanagira
Paste tense	31.70	Undu	Illai
Azhapadai		Niruthi	-
Yeval	34.73	Undu	Illai
Ugara serkai		Un, thin, sel	unnu, thinnu, sellu
Thozhipeyar	26.67	Thal	kiradhu (seygiradhu)
Yedir marai	35.75	Arka	Aathey
Yeval		(seyyarga)	(Seyyathey)

Two manuscript copies of the grammar of High-Tamil are kept in the collection of the Bibliothèque Nationale in Paris and they differ considerably from each other, while the second copy is shorter but a good deal more correct. We find one more copy of the manuscript complete but less exact.[59] A manuscript copy is also available at the Madurai Province Jesuit Archives, Shenbaganur, India. Leon Besse's edition was made from this manuscript which was then he kept in the Residence of the Jesuits at Tiruchirapalli. It was a booklet of 103 pages, worn out by time, usage, and the Indian climate. This manuscript was without title, preface, or tables. After having compared it with Babington's translation we are convinced that it was substantially the same work. A friend in Paris had copied out the Preface, from one of the manuscripts of the Bibliothèque Nationale and the manuscript was transcribed for the press by a young religious of the Society of Jesus.[60]

A copy of Beschi's grammar of High-Tamil preserved at the British Museum, London, is in excellent condition.[61] The title-page of the work reads *Grammatica Latino Tamulica ubi de elegantiori linguae Tamulicae dialecto [Senthamizh] dicta, cui adduntur Tamulicae poesios rudimenta. Autore R. R. C. J. Beschio Soc. is Jesu. Missionario in Regno Madurensi.* This copy was made in 1814. It seems to be the work of two different copyists. The first part up to folio 76 shows a round and in a generous writing. The second part is written in a thin, pointed, and somewhat crowded hand. The first copyist seems to have been more careful. There are throughout the manuscript a number of spelling mistakes, occasional omissions of words and short passages, interpolations or changes of words in the sentences. Some of the actual mistakes have been corrected in black ink, and additional corrections have been made with pencil. The person who made the latter must have had a thorough knowledge of classical Tamil literature, for Babington had written in the introduction to his translation that Beschi in his preface, told us, that he has not quoted authorities for his examples, because the names of the authors even of the most celebrated works were unknown.[62] In this British Museum copy most of the quotations from Tamil classics bear footnotes in pencil giving chapter and verse of the work from which they have been taken.

The Latin text of the manuscript seems to be almost identical with the one preserved at the Madurai Province Jesuit Archives, given in Leon Besse edition. We find the same number of chapters (145), though in the British Museum copy they have occasionally been numbered wrongly, and in a few cases the copyist forgot to mark the chapters. Beschi's Latin preface appears in full, corresponding more or less to the one copied from the Bibliothèque Nationale, Paris manuscript, although a short passage towards the end of the preface is missing and the very last part of it is slightly different. There are also differences in punctuation throughout the manuscript and some words given in Tamil characters in Leon Besse's edition are written in Roman script and vice versa. In chapter iii, paragraph 5, no. 69, a short passage which appears in Leon Besse edition is given but has later been crossed out with black ink. Paragraph I, in chapter IV (De Arte Tamulica Poeseos) of part II, which in the Leon Besse edition is titled 'De Dictione poetica' is here called 'De Phrasi Poeticas'.

An English translation of the Latin text was published by Benjamin Guy Babington titled 'A Grammar of the High Dialect of the Tamil language' in 1822. In his preface Babington gives useful information. European students in the colleges learnt Tamil language mainly through Beschi's grammar written in Latin. Tamil language was taught as a subject to Europeans in colleges. Beschi's Tamil grammar in Latin was very popular and it was circulated in manuscript form before it got printed. Kodunthamizh was translated into English first and later senthamizh was translated into English language.

Babington in his translation had added several things which were not found in Beschi's original in Latin version. The letter 'a' gets saariyai and called 'agaram' is mentioned in Latin original by Beschi. But Babington while mentioning this added Thiruvalluvar had expressed 'agaramuthala yeluthellam' and gives examples. Further Beschi while explaining the 'ayutha' letter uses the terms, 'ithu and ihuthu'. But Babington uses the letter 'athu and ahuthu' and further cites the kural 'Anbin vazhiya thuir inilai yahuthillark kenbu thol portha udambu'. Thus there are several insertions that had been made by Babington.

Thonnul Vizhlakkam: The Grammar for Literary Tamil in Tamil by Beschi, 1731

Beschi wrote a treatise in Tamil titled Thonnul Vizhlakkam in 1731 and it is a grammar work that Beschi had added. This Thonnul Vizhlakkam is a prosody which he composed. It treats literary Tamil (senthamizh) but it consists of a summary of the five parts of the traditional Tamil grammatical treatises, written in verse with a commentary by Beschi. In the invocation song he mentioned the six attributes of God such as, he is of himself the lord of all; he is eternal; he is immaterial; he manifests himself in everlasting and universal goodness; he pervades all space and; he is the first cause of all.[63]

Beschi's fame rests partly on brilliant linguistic achievements, partly on a rare poetic sensitiveness that enabled him to penetrate the complex thoughts and ornate techniques of Tamil poetry and produce some highly original works in so alien a medium. No other European had acquired such a mastery over Tamil, especially over its classical Tamil.

Thonnul Vizhlakkam followed traditional Tamil grammars especially Nannul in its overall organization of five divisions, sol (words), porul, (meaning) ezhuthu (letters), yappu (prosody) and ani (rhetoric), but it introduced new features as well, such as extensive use of prose and a section on the four Hindu goals of life. In a difficult exercise of synthesis, Beschi also reduced the early thousand grammatical rules (in most traditional grammars) to just 370 *suththirams*.

Beschi in his Thonnul Vizhlakkam had largely used Nannul suttrams and the comparative details found both in Nannul and Thonnul Vizhlakkam would be interesting to mention here.

Nannul	*Thonnul Vizhlakkam*
309 - Ittrin	74 - Ittrin
199 - Munnathin	97 - Kurugittru
186 - Kavvottru	99 - Kavvodu
307 - Ural	72 - Varal
347 - Irattal	68 - Iratta

Further, he had also used two or more suttrams of nannul and compressed them into a single suttram and even one suttram was converted into parts and Beschi employed them in Thonnul Vizhlakkam. However, he had fully used the same suttram as in Nannual in his work Thonnul Vizhlakkam as follows.

Nannul	*Thonnul Vizhlakkam*
(67) 350	102. 3
168	116. 3.4
174	98. 3-5
136	92. 2-5
300 (4th suththiram Aa)	67 (4th adi - serntha)
305 (2nd suththiram Immup)	70 (3rd adi ye: um)

It must be said that the traditional grammars typically only framed rules and adduced examples to illustrate but Beschi had commented upon them in prose.[64] Thirukural had been extensively cited by Beschi in Thonnul Vizhlakkam and the details are as follows:

Suththiram number	*Kural number*	
18	653	Othal vendumozhi
18	1200	Uraaark kuruno uraippay

18	15	Keduppathuung kettaarkku
18	1115	Anichappu kaalkazhlaiyaal
18	1263	Uran nasaiee ullanthunaiyaaga
19	66	Kuzhlalinithu yaazhlinithenba
60	934	Sirumai pala seythu
60	964	Thaliyin nilantha mayiranaiyar
124	502	Kudipiranthu kuttrathin neengi
149	34	Manathukan maasila
155	726	Vaaloden vankanna
160	146	Pagai paava machcham
160	320	Noyella noy seythaar
161	426	Yevva thuraiva thulaga
162	272	Vaanuyar thottramavan
205	3	Malar misai yeginaan
208	254	Arulalla thiyaathenir kollaamai
213	350	Pattruga pattrattaan pattrinai
214	74	Anbinoo maarva mudamai
214	114	Thakkaar thavila renba
214	292	Poymayium vaaymay idaththa
214	792	Aayanthaayunthu kozhlaathaan
220	207	Yenai pagai uttaaru muyvar
220	426	Yevva thuraivaga thulaga
221	36	Anrarivaa mennaa tharang cheyha
340	872	Villycr uzhavar pagai kozhlinum
341	408	Nallar katpatta varumai

The Translation of the Grammar for Literary Tamil into Latin by Beschi, 1732

Thonnul Vizhlakkam written in Tamil was translated by Beschi into Latin for Europeans in 1732. He gave in Latin the title Clavis Humaniorum Litterarum Sublimioris Tamulici Idiomatis. This adapted version in Latin was called 'Key to the refined literature of the Higher Tamil language'.[65] A study and examination of the Clavis reveals that Beschi mentioned kuttriyalugaram in ezhuth ilakkanam and one can see also agu peyar in sol ilakkanam Beschi mentioned in porul ilakkanam the kurinji thinai, and in yapp ilakkanam one can find monai and in ani ilakkanam one can notice uruvagam. He also quotes from Thembhavani his epic for explaining uruvaga ani.

Beschi had extensively cited Thirukural in the grammar for literary Tamil while translating into Latin and the details are as follows:

Paragraph number	*Kural number*	
1.13	2	Kattrathanaalaaya payanenkol vaalariva
2.41	57	Sirai kaakung kaapevan
2.41	127	Yahaava raayinu naakaakka
2.41	763	Olithakkaalenna muvari yelipagai
2.48	496	Kadal odaa kaalva nedunther
2.50	439	Viyavaraka vengnnaanrun thannai
2.55	660	Slathaar porul seythe maarthal
2.56	319	Pirarkinna murpagar cheyyir
2.59	450	Pallar pagaikozhlarir patheduththa
2.59	1	Agara muthala vezhuthellaam
5.3	1038	Yerinu nanraaleru viduthal
5.15	452	Nilathiyalbaa neer thirinthattraagu
5.15	109	Konranna vinna seyinunumavar
5.41	1080	Yettirkuriyar kayavaron

Tamil Grammar in Latin by Onorato de Udine, 1793

In spite of the fact that Tamil grammar was composed in Latin by Ziegenbalg, Beschi and Christoph Theodosius Walther we find Onorato de Udine a Capuchin who worked in Madras who learnt Tamil had composed a grammar of Tamil in Latin. He had completed it in 1793. This 'Gramatica Tamulica' is preserved at the Biblioteca Nazionale Centrale, Vittorio Emmanule II, Roma.[66] It is very much an adaptation of Beschi's earlier works but giving numerous examples for daily use.

Tamil Grammar in French: Observation of Dupuis and Mousset, 1855

Louis Savinien Dupuis and Louis Marie Mousset the two MEP missionaries had studied Tamil grammar and they mention that in Tamil there is only one conjugation for all verbs, with only three tenses, past, present and future, independent of the infinitive. The conditional has no person; the same word serves for all, and includes the meaning of the conjunction if. Sanskrit, however like Latin, has several conjugations, each with many moods and tenses.[67]

They also mention that Tamil has no conjugation or special inflections for the passive. The conjugation of the intransitive verb padugirathu, whose root and meaning are curiously close to the Latin, pati (suffer) is joined to the infinitive of active verbs, and sometimes even of neutral verbs to form the passive. In Sanskrit, such verbs are conjugated separately. Tamil has a particular way of expressing the negative of verbs; a special tense, with all persons, can express this in one word. The negative verb has an imperative, participles and gerundives. Thus 'Arreyn' means I am not. There is nothing similar in Sanskrit.

Unlike Sanskrit or European languages, Tamil has no who, which or other relative pronouns. These pronouns are understood in the participles themselves, and just as with the locative adverb where, so are prepositions governing relatives, such as in which, with which etc. Tamil has no dual such as have Sanskrit and Greek. But in the ordinary language, there are honorific suffixes, that is, to be formal and respectful to one's addressee. These endings serve as honorifics in ordinary Tamil, and as plurals in elevated Tamil, which like Latin always uses the singular when speaking to or about a single person.

Although nouns are declined in Tamil, there is really only one declension, and the case suffixes are the same in the singular and the plural. It is the changes at the ends of certain nouns, or the insertion of two letters before the suffix, which distinguishes the plural. Sanskrit has several declensions and the singular and plural endings are different. In Tamil, adjectives are indeclinable, and the way of forming them is quite particular. In Sanskrit, on the other hand, adjectives are declines and are formed differently.[68]

According to Dupuis and Mousset all of this proves quite clearly that Tamil does not come from Sanskrit; that consequently the Tamil language, and the people who speak it, existed before the arrival or invasion of the Brahmins. Many writers agree that the reign of Tamil goes back a long way in India, one can find traces of it in the remotest times. The Brahmin invasion pushed it southwards in the peninsula, where the people and the country have never lost it. For a long time it flourished at Madurai, where, under the Indian kings, it was cultivated by a type of academy. It has always preserved its claims of being more ancient and vested its rival with foreignness by calling Sanskrit and words derived from it vada mozhi (language or words from the north). In addition, the Brahmins and Sanskrit which they imported into this region are denoted by the words Aariyar or Aariya, which obviously come from the word Aariyam the name of an area between the Himalayas and the Vindhya range (unless one derives it from Arya, the name of a province of Persia). Whatever it may be, it is generally accepted, especially by scholars that the Brahmins came from, or through, the north of India. Very probably they implanted most of the teachings and the idolatrous practices of India, her mythology and current polytheism etc at the same time as they introduced the Sanskrit words used to express these concepts. Tamil has always preserved, even after the Brahmins ascendance in India, its distinct national character, and one can presume that it was well cultivated, and that it continued to be the language of not only the poorest people but also of scholars and the great majority of the nation; otherwise it would have disappeared or merged with the language of the rulers, such as happened in so many countries. A host of expressions contrary to polytheism, others which support Buddhism and everything related to it, and which have been retained in the language; the Jains and the Samani, quite a number of whom still live in the region; historical facts; ancient coins or rather traces of antiquity collected by scholars—all seem to prove quite clearly that one kind of Buddhism, or another sect closer to the true religion, and very different from Brahmanism, was once predominant in India. Wars and cruel persecutions nearly wiped it out, leaving its rival which roused passions to a greater degree—in a superior position. There are even some authors who have gathered vestiges of antiquity to claim that the reign of Brahmanism and its worship of idols, i.e., very much later than Christianity. All of this categorically denies the great antiquity which the Brahmins have sought to give their mythology and the books they hold sacred, and which the Buddhists have rejected with such aversion, as have the Christians, enlightened by the truth.[69]

If Tamil does not come from Sanskrit, where does it come from? Is its origin the same as that of Pali? Dupuis and Mousset mention that they are unable to say. Is it truly a mother-language? They state that they would not go so far as to assert this. It has many descendants, but its origin is lost in the mists of time and they are not able to determine the epoch. It is at least certain that it must be later than the confusion of languages which occurred at the Tower of Babel, and consequently, much later the Flood, whose tradition is preserved in this region as elsewhere. The era of kali Yuga the only one of the four Indian yugas which one can consider as not mythical-although not wholly so—goes back to three years after the flood, following the chronology of the Septuagint before the patriarch Abraham, and that of the Hebrews after him. One thing worth noting is that Tamil, Pali, Sanskrit, Tibetan, Thai and several other languages of India and of the neighbouring countries classify their letters in the same order, which shows that they must have something common in their origins.[70]

Missionaries wrote Tamil grammars for beginner learners who arrived from Europe who were trained in the Graeco-Latin tradition. Otto Zwartjes opines that missionary grammarians-lexicographers had worked as translators.[71] This view is called into question because translating grammar into Tamil was not done. Further, they compared and contrasted the grammar parts and had not manipulated either the grammar or the lexicon of Tamil, but tried their best to make it simple to the European learners.

To conclude it may be said that the Tamil grammatical works of the missionaries had become the main learning tools to learn both the spoken and literary Tamil for the European language speakers. It is important to know how the native grammatical works in poetry were made available in prose for the Europeans. In fact it is true that Tholkaapiyam the earliest Tamil grammar work gives elaborate rules of language use and theories of poetry, prosody and rhetoric. It prescribes well codified rules for poetical conventions distribution of phonemes, linguistic expressions and nativization of borrowed lexical items or elements. Intellectual acuteness was developed by Tholkaapiyar through the study of Tamil grammar. It is applicable only to him. Nannular another author of Tamil grammar wrote pathaviyal in between ezhuthiyal and punariyal and it is significant from the view point of doctrine or principle. Following the footsteps all others wrote Tamil grammar like the earlier Tamil grammarians by improving which is found in their works. Europeans were informed that there were a number of commentators of Tholkaapiyar's grammar in Tamil and mention was made of Ilampuranar, (probably Tenth Century) otherwise known as urai aciriyar. He was said to be the earliest known commentator of Tholkaappiyam. Others like Senavarayar, Dheyvachilaiyar and Nachinarkiniyar (whose periods are not exactly known) were the other well known commentators of Tholkaapiyam. It was also told that there is another work of Tamil grammar called Vira Chozhliyum by Buddhamithiranar (Eleventh Century). Perunthevanar (Eleventh and Twelfth Centuries) made a commentary on Vira Chozhliyam. Further Kunavirapandithar, (Thirteenth Century) the author of Neminatham, and Bhavananthi (Thirteenth Century), the author of Nannul also produced works on Tamil grammar. Mayilainathar (Fourteenth Century) was the earliest commentator of Nannul. All these fund of information must have highly perplexed the missionaries as to study which grammatical treatise in poetry that would suit their requirements.

It may be appropriate to ask as to how the Europeans felt the necessity to write a grammar in Tamil for the European audience. As they were fully aware that grammar composed in most intricate poetry would not be helpful and useful, and served no purpose to learn Tamil. Hence they began to write Tamil grammar and it rules in prose and particularly in the various Europeans languages such as Portuguese, German, Latin and French which would serve as a language tool. It may be useful to note the processes through which the language and grammar in European languages evolved and developed to meet the requirements of that period. Archaic Portuguese was much used in speech that consisted of popular Latin usages from the twelfth century to the sixteenth century. Later Portuguese language entered a definite and regular phase by taking ample and opulent forms which were shaped by the help of the early grammarian Fernão de Oliveira who wrote Gramattica da Lingoagem Portuguesa in 1536. João de Barros who wrote Grammatica da Lingua Portuguesa in 1540 was himself known to Fr Anrique Anriquez and therefore he used and mentioned the same salient features while composing the Tamil grammar in Portuguese. It is to be noted that the Portuguese language was just then coming of age leaving behind the archaisms of its infancy, Tamil with its age old traditional grammar which had already attained adulthood with its rich vocabulary and literary elegances attracted the attention of the Portuguese missionaries prompting them to learn the oriental language with seriousness and dedication. Portuguese missionaries pointed out that the nouns in Portuguese language had no case forms, though nouns were distinguished only between singular and plural which was not the case in Tamil. Further, the second and the third case, fused into a single one in Portuguese language at that time is mentioned while it was not so in classical Latin. Based on their experience with Portuguese language, they pointed out the salient features of Tamil grammar. They informed that all nouns in Tamil denoting inanimate substances, irrational beings are of the neuter gender.

A close study of the manuscripts of the various Tamil grammars in Portuguese reveals certain similarities too. The infinitive verb visuvasiththal (to believe) is used for conjugation by Anrique Anriquez in folios 33r-35r and the same infinitive is again used by Philippus Baldeaus who wrote Tamil grammar in Dutch. In my opinion this is an important word from the point of view of Christian theology, because the whole edifice of Christianity is built on the bed rock of faith. This will also perhaps be the most important word in the Tamil Christian vocabulary of any believer of that time. The days of agnosticism were still far away. Similarly the noun pandam (commodity) is used in the Anriquez manuscript on folio 17r and the same noun is again found mentioned in the manuscript of Paris in folios 5-6 and this word is an important word from the point of view of the then emerging international trade conducted between Europe and Tamil country at that time. So also with the word raasaa (king) used in the manuscript of Anriquez on folio 18r, and the same noun is again used in the Paris manuscript copy on folios 5-6 and 8 which is important from the point of view of the political and diplomatic relations.

Anriquez wrote grammar on the spoken Tamil. His Arte de Lingua Malavar served as the base for others. Philippus Baldaeus in 1672 had used and repeated the material of Anriquez and we find the same present participle with the word vichivasikiravan in his writings on Tamil grammar. Further, Balthasar da Costa also used the term visuvasi, of Anriquez as present participle. An examination of the texts of the Tamil grammars in Portuguese composed by these various missionaries, we find new Tamil words relating to Christian expressions. For instance, the verb, 'to believe' is mainly used for conjugation in all their texts. Other examples of some simple sentences include such as, 'If Peter believes in God, I will treat him well', 'It is easier to believe in idols than in God', 'To believe in idols is as much as to believe in demons' found in the Anriquez manuscript on folios 42v, 44r, 44 and 61r.

Portuguese missionaries were pioneers in the field of Tamil grammar writing in prose from the mid-sixteenth century to the end of the eighteenth century and ushered the Tamil language, the knowledge of which, until then held as the exclusive preserve of the 'chosen and privileged few' into the modern era by triggering the process of its modernization and accelerating the process of its democratization. The concept of humanism rising to prominence in Portugal at that time mainly had also an impact in the process of democratization of Tamil and it began to get into the Tamil stream.

While studying Tamil language to speak and write the missionaries doing so they learnt Tamil grammar. These missionaries noted the differences between European languages and pointed out Tamil language has no article, no relative pronoun no dual number. In the declension of its nouns, the conjugations of its verbs and the arrangement of its sentences it more resembled the Latin. These works remained in manuscript; each newly arrived missionary making a copy of them for his own use. In course of time they were enlarged and improved. A new impetus was thus given to the study of Tamil.

Notes

1. Jose Wicki, *Documenta Indica* (hereafter DI), 18 vols., Roma, 1948-1988, vol. I, pp. 429-30, see the letter of Paulo della Valle from Vembar dated 24 December 1548 to Luis de Gra; Antonio Franco, *Synopsis Annualium Societatis Jesu in*

Lusitania ab annis 1540 usque ad annum 1725, Haeredum, Anno MDCCXXVI, p. 25.

2. DI, vol. I, pp. 581-2, see the letter written from Punnaikayal dated 21 November 1549; Raphael Ricci, *Selectae Indicarum Epistolae nunc Primum Editae*, Florentiae, 1887, p. 93.
3. Ibid., vol. II, pp. 388-9, see the letter of Anrique Anriquez from Punnaikayal dated 5 November 1552; See also, vol. I, document no. 85, p. 581.
4. Ibid., vol. I, pp. 286-8, 488-9.
5. Biblioteca Nacional de Lisboã (hereafter BNL), Portugal, MSS Reservados, Codice 3141, fl. 155v.
6. Antonio da Silva Rego, *Documentação Para a Historia das Missões do Padroado Portugues do Oriente*, 12 vols., Lisboã, 1947-58, vol. I, p. 153.
7. DI, vol. I, pp. 581-2, see the letter written from Punnaikayal dated 21 November 1549.
8. Ibid., vol. I, p. 584.
9. Ibid., vol. II, p. 166.
10. Ibid., vol. IV, p. 127.
11. Ibid., vol. IV, p. 159.
12. Ibid., vol. VI, p. 396, See the letter written from Kollam dated 27 January 1566; see also, vol. VI, p. 751.
13. Ibid., vol. VII, p. 442.
14. BNL, MSS Reservados, Fundo Geral, Arte Malauar, no. 3141, fl.5-6v; Hans J. Vermeer and Angelika Morath, *The First European Tamil Grammar*, Heidelberg, 1982; J. Hein and V.S. Rajam, *The Earliest Missionary Grammar of Tamil: Fr. Henriques' Arte da Lingua Malabar: Translation, History and Analysis*, Cambridge, Harvard University Press, 2013; Michelle Zaleski, 'Beyond Words: Missionary Grammars and the Construction of Language in Tamil Country', *Encounters between Jesuits and Protestants in Asia and the Americas*, Leiden, 2018, pp. 159-170.
15. DI, vol. VII, p.174.
16. Ibid., vol. I, pp. 285-7 (31 October 1548); vol. I, pp. 581-582 (31 October 1548 and 21 November 1549).
17. Ibid., vol. VII, pp. 437-443.
18. St. John's Abbey and University College Ville, Minnesota, A Arte da Lingua Malabar, microfilm no. 790; Xavier Thaninayagam, 'Tamil Manuscripts in European Libraries', *Tamil Culture*, vol. III. October 1954, pp. 210-88; Michael Manickam, *On Describing a Non-Indo-European Language in Traditional Indo-European Grammatical Categories*, Unpublished Master's Thesis, Hartford Seminary Foundation, Hartford, 1968; H. Jeanne, 'Father Henriques' Grammar of Spoken Tamil' *Indian Church History Review*, Vol. 11, 1977, pp.127-57; Cristina Muru, *Missionari Portoghesi in India nei Secoli XVIe XVIIe: L'Arte della Lingua Tamil: Studio Comparati di Alcuni Manoscritti*, Roma, 2010; Antonio Franco, *Synopsis Annualium Soceitatis Jesu in Lusitana ab annis 1540 usque ad annum 1725*, Haeredum, MDLCCXXVI, see, Index materiam dequibus tracarunt Lusitane Provinciae scriptores ab initio Societatis ad annum 1724, p. 520.

19. BNL, MSS Reservados, Fundo Geral, Arte Malauar, no. 3141, fls. 23.
20. Ibid., fls. 5-6.
21. Ibid., fl. 5v.
22. S. Jeyaseela Stephen, *Portuguese-Tamil Grammar: Modernization and Democratization of Tamil*, Pondicherry, 2001.
23. Hans J. Vermeer, The First European Tamil, pp. ix-xx. See also, Gregory James, 'The Terminology of Declension in Early Missionary Grammars of Tamil', in *Missionary Linguistics III/Linguistica Misionera III, Morphology and Syntax: Selected Papers from the Third and Fourth International Conferences on Missionary Linguistics, Hong Kong/Macau, 12-15 March 2005, Valladolid, 8-11 March 2006*, eds., Otto Zwartjes, Gregory James and Emilio Ridruejo, Amsterdam, 2007, pp.167–190.
24. Ibid, p. xix.
25. Ibid., fl. 6v.
26. DI, vol. I, p. 287.
27. DI, vol. VI, p. 396.
28. State Central Library (hereafter SCL), Goa, MS. 35, Pe. Henrique Henriques, Vocabulario Lusitano-Tamulico Seguido d'uma Arte Tamulica, séc. XVI.
29. S. Rajamanickam, *The First Oriental Scholar*, Palayamkottai, 1972, p. 267.
30. Institutio Arquivo Nacionais/Torre do Tombo (hereafter IANTT), Lisboã, MSS *Documentos Remetidos da India*, Livro 44, fl. 73; Livro 45, fl. 67.
31. SCL, Goa, MS. 49, Arte Tamulica composta pello Pe. Balthazar da Costa da Comp. a de IESV.
32. Staats-und Universitatsbibliotheek Hamburg Carl von Ossietzy, Cod. Orient 283, fl. 70, Arte Tamul Sive Instituto Gramatica Linguae Malabaricae Sum Philippi Baldej. UDM in Regno Jaffnapatam, 1659, Philippi Baldeyusu.
33. Cristina Muru, 'Gaspar de Aguilar: A Banished Genius' in Anand Amaladass and Ines G. Zupanvo, eds., *Intercultural Encounters and the Jesuit Mission in South Asia, 16th -18th Centuries*, Bangalore, 2014, pp. 353-389.
34. J.A.B. Van Buiteen and P. C. Ganesh Sundaram, 'A 17th Century Dutch Grammar of Tamil', *Bulletin of the Deccan College Research Institute*, Poona, vol. XIV, no. 3. December 1952, pp. 168-82; T.P. Meenakshi Sundaram, 'Tamil Letters and Sounds of the Seventeenth Century', *Collected Papers of Prof. T. P. M, Sixty-first Birthday Commemoration Volume*, Annamalai Nagar, 1961, pp. 225-31.
35. Archives Françaises de la Compagnie de Jésus (hereafter AFCJ), Vanves, Hauts-de-Seine, MSS Index 3; S. Jeyaseela Stephen, 'Indian Manuscripts Sent from Chandernagore and Pondicherry (AD 17141735): Role of the French East India Company and Jesuit Missionaries', *Journal of the Oriental Institute*, vol. 50, no. 1-4, September-June 2000-2001, pp. 245-248.
36. Bibliothèque Nationale, Paris (hereafter BNP), Rue Richelieu, MSS Indien, no. 188.
37. S. Jeyaseela Stephen, *Portuguese in the Tamil Coast: Historical Explorations in Commerce and Culture, 1507-1749*, Pondicherry, 1998, p. 351.

38. SCL, Goa, MS. 49, Arte Tamulica composta pello Pe. Balthazar da Costa da Comp. a de IESV, copy dated 1731. There is a catechism in Tamil which is included after folio 151; See also SCL, Goa, MS 16, Balthazar da Costa, Arte Tamulica.
39. British Library (hereafter BL), London, Sloane Manuscripts, no. 3003, *Grammatica Linguae Malbaricae, Verbis Lusitanicis Explicata*. The copy of Balthasar da Costa's grammar manuscript is dated 25 June 1685.
40. Ibid., fls. 23-55.
41. Historical Archives of Goa (hereafter HAG), Panaji, MSS Monçōes do Reino (hereafter MR), livro 177-A, fl. 37.
42. J. H. Cunha Rivara, *A Jurisdiçaõ Diocesana do Bispado de S. Thomé de Meliapor nas Possesções Francezas e Inglezas*, Nova Goa, 1867, p. 108.
43. Leon Besse, *La Mission du Maduré: Historique de ses Pangous*, Trichnopoly, 1914, pp. 535, 539.
44. HAG, MR, livro 182, fl. 36.
45. Arquivo Historico Ultramarino (hereafter AHU), Lisboã, MSS Documentos Avulsos, India Maço, 187.
46. BL, Sloane Manuscript, 3003, fls.13.
47. SCL, Goa, MS 34, Pe. Antão de Proença, Vocabulário Tamulico- Luzitano Precedido de Uma Arte Abreviada em Tamul e Seguido de Termos Poéticos Anno 1670.
48. BNP, MSS Indien, no. 221 fls.1-197. This copy was made by Anquetil Duperron in 1750. See also MSS Indien, no. 222.
49. Philippus Baldaeus, *Nauwkeurige beschryvinge van Malabar en Choromandel, derhalver aangrenzende ryken, en het machtige eyland Ceylon, Nevens een omstandige en grondigh doorzochte ontdekking en wederlegginge van de afgoderye der Oost-Indische heydenen*, 3 vols, Amsterdam, 1672, pp. 186-90.
50. Ibid., pp.191-2.
51. BL, Sloane Manuscript, 3003, fls.23-55.
52. T.P. Meenakshi Sundaram, 'A Study of Baldaeus's Grammar', *Indian Linguistics*, vol. 1, 1958, pp. 7-19.
53. Arno Lehmann, ed., *Alte Briefe aus Indien: Unveröffentlichte Briefe von Bartholomäus Ziegenbalg, 1706–1719*, Berlin, 1957, pp. 42-3, 59.
54. Ibid., p. 39.
55. Ibid., p. 56.
56. Det Kongelige Bibliotek (hereafter KB), Kobenhavn, NY.Kgl.Saml.589C, Christoph Theodosius Walther, Bibliotheca Tamulica, consistens in recensiane librum nostrorum mscr-torum ad cognoscendam et linguam & res Tamulicas inserventium 1731, p. 151.
57. Archiv der Franckesche Stiftungen (hereafter AFSt), Halle, MSS M1 C1:66, fl. 5.
58. BL, London, MSS Or.13044, fls. 2-4; Church Mission Society Library and Archives, Birmingham, *A Grammar of the High Dialect of the Tamil Language Termed Shen-Tamil to which is added an Introduction to Tamil Poetry: Translated*

from the Original Latin by Benjamin Guy Babington, Printed at the College Press, Madras, 1822; In his introduction Babington mentioned that it was the case with the copy from which the version was made, and it was only by the collation of several texts, that the faulty passages have been amended or explained. In 1917 Leon Besse published Babington's translation together with the Latin text '*A Grammar of High-Tamil*', St. Joseph's Industrial School Press, Trichinopoly, 1917.

59. Julien Vinson, 'Note Supplementaire sur les Oeuvres Tamoules du P. Beschi', *Revue de Linguistitue et de Philologie Comparée*, vol. XXXIII, 1900, pp. 198-200.
60. Leon Besse, *Father Beschi : His Times and his Writings*, Trichnopoly, 1918, p. 217.
61. BL, London, MSS Or.13044, fls.1-124.
62. C. J. Beschi, A Grammar of the Higher Dialect of the Tamil Language termed Shen-Tamil, p. iii.
63. Srinivasa Raghavacharyar and G. Mackenzie Cobban, ed., *Ainthilakkana Thonnul Vilakkam: Mulamum Uraiyum*, Madras, 1891.
64. K. Meenakshi Sundaram, *The Contribution of European Scholars to Tamil*, Madras, 1974, p.169; Jean-Luc Chevillard, 'Beschi, grammairien du tamoul, et l'origine de la notion de verbe appellatif'. *Bulletin de L'Ecole française d'Extrême-Orient*, Tome 79, No. 1, 1992, pp. 77-88.
65. BL, London, Or. no.14172.f.17, *CLAVIS Humaniorum Litterarum Sublimioris Tamulici Idiomatis, Auctroe R.P. Constantio Josepho Beschio Societatis Iesu, In Madurensis regno Missionario*, Evangelical Lutheran Mission Press, Tranquebar, 1876. It was printed for A. Burnell, by the Evangelical Lutheran Mission Press in Tranquebar. The preface of the book written by A. Burnell from Thanjavur in 1876 gives interesting information. Beschi' treatise on the so-called High-Tamil and on Tamil poetry and rhetoric is here printed for the first time from a manuscript corrected by the author himself. The Rev. K. Ihlefeld has undertaken the very laborious task of editing it precisely as it stands in the manuscript, for corrections could hardly be allowed in this case. The Tamil quotations have been carefully revised according to the text given in Beschi's Thonnul Vizhlakkam. The manuscript of Clavis was purchased on the sale of Babington's library and he got it from Mr. F.W. Ellis who had a search made, about the beginning of the nineteenth century, for Beschi's manuscripts.
66. Bibliotheca Nazionale Centrale di Roma, Vittorio Emmanule II, (hereafter BNCR-VE) MSS Varia 53 (598), fls.1-85.
67. Société Archeologique de Sens (hereafter SAS), *Dictionnaire Tamoul-Français par deux missionaries Apostoliques de la Congregation des Missions-Etrangères*, Pondichéry, 1855, see the preface, pp. viii-ix.
68. Ibid., p. xi.
69. Ibid., p. xi-xii
70. Ibid.
71. Otto Zwartjes, *Portuguese Missionary Grammars in Asia, Africa and Brazil, 1550-1800*, Amsterdam, 2011, p. 11.

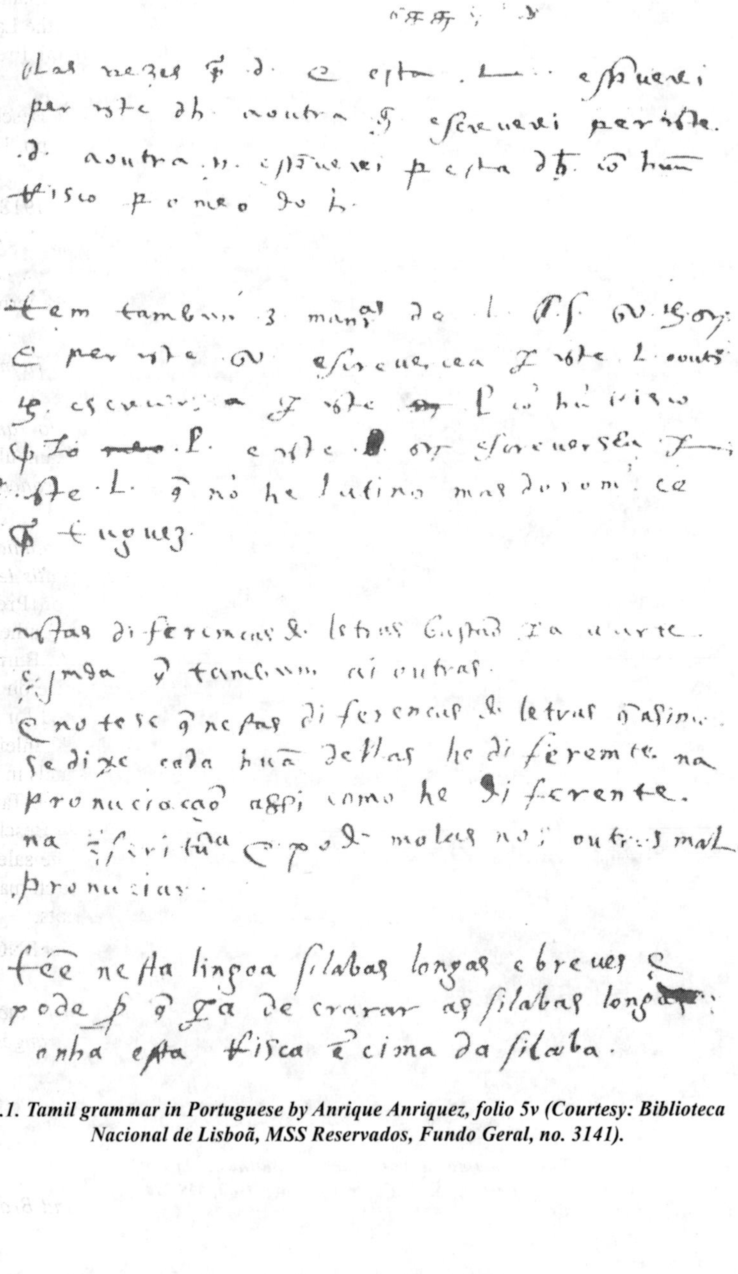

3.1. Tamil grammar in Portuguese by Anrique Anriquez, folio 5v (Courtesy: Biblioteca Nacional de Lisboã, MSS Reservados, Fundo Geral, no. 3141).

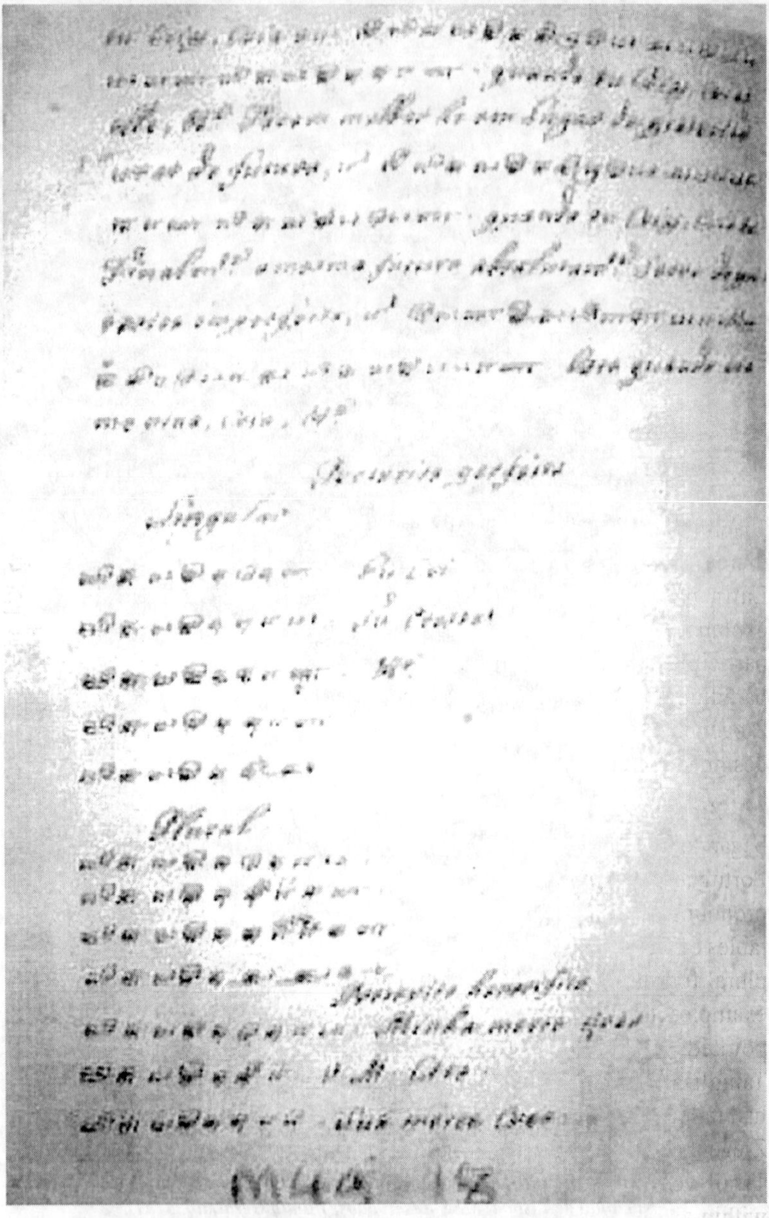

3.2. Tamil grammar in Portuguese by Balthasar da Costa, folio 18 (Courtesy: State Central Library, Goa, MS M49) .

4

Printing Tamil Grammar in Latin and English for the Western Audience 1716-1843

Bartholomäus Ziegenbalg the protestant missionary came to know Joachim Lange's Latin grammar published in Germany in 1707. He thereafter began to use and structure his Latin-Tamil grammar according to Latin patterns and rules. Ziegenbalg's grammar prepared contains explanations about letters, pronunciation, adjectives, pronouns, verbs, participles and syntax. He gave examples and presented in tables and paradigms to explain and reinforce grammar rules. It should be noted that Ziegenbalg thought that his European readers would understand a grammar designed according to a Latin pattern rather than to a Tamil pattern.[1]

Ziegenbalg who was aware of Philippus Baldaeus's Tamil grammar began to use.[2] He extensively used Balthasar da Costa's Tamil grammar in Portuguese which he found convenient. We find the declensions of the pronouns and their conjugations in chapter five that are identical.[3] The tables of verbs, including the affirmative, honorific, negative, singular and plural forms, tense forms, imperatives and moods are also identical.[4] The examples and their meanings agree word to word. Hence Daniel Jeyaraj pointed out that Ziegenbalg had translated Balthasar da Costa's Arte Tamulica the Portuguese text of Tamil grammar into Latin and Ziegenbalg had failed to acknowledge it. However, he did some work on his own since Ziegenbalg had elaborated the further examples and added several list of Tamil words for further exercise.[5] Thus he was able to compile the grammar within a short period of three months commencing on 15 February 1715 during the voyage from Cape of Good Hope to Ascension Island in the South Atlantic Ocean.[6] On 5 June 1715 Ziegenbalg sent a letter to Professor Michaelis that he had finished writing the Latin-Tamil grammar and it should be printed in Halle in Germany. He sent the manuscript through Polycarp Jordan and requested Michaelis to proof read the Latin and if necessary to improve its style. Michaleis replied on 6 August 1715 that Jordan met Heinrich Julius Elers (1667-1728) in Berlin and handed over the manuscript.[7] The simple reason for not able to print the Latin-Tamil grammar book in Tranquebar had been stated by Ziegenbalg because the Tamil letters made were proportionately much bigger than the Latin letters. Further, the Tamil letters and the Latin letters could not be put together and got printed in a book of bi-lingual nature at Tranquebar.[8]

Printing Latin-Tamil grammar was realized by the Protestant missionaries. This was felt chiefly important for it would help the missionaries later to teach the tenets of the religion to the natives in their language. Hence, missionaries in Europe were driven to print the spoken Tamil grammar, master Tamil language and also in the near future to develop Christian Tamil prose. It was not found practical to ask every missionary to make his own arrangement to learn the language. Tamil grammar book printing was felt necessary so that it would facilitate and help in a different and better way to learn the language. There was no other way left out in the beginning to learn Tamil for the missionaries before they would come to Tamil country for apostolic work. These circumstances had prompted some of the very ardent Protestant missionaries to take immediate steps to print Tamil grammar.

Tamil Grammar in Latin by Ziegenbalg: Printing at Halle in Germany in 1716

Bartholomäus Ziegenbalg's Grammatica Damulica was published in the missionary printing press at Halle in Germany in 1716. It has xiv+128 pages.[9] It is titled Tamil grammar which through a variety of paradigms, rules and necessary vocabulary demonstrates the shortest route to learn the Tamil language, which is also known as Malabarian that is used by the East Indians and that until this time was unknown in Europe, may easily be learned by those who at this time wish to use it and desire to lead those peoples from the worship of images to the worship of the true God and to eternal life revealed by the Gospel of Christ. Prepared en route to Europe, in a Danish ship (named Friedrich IV), by Bartholomäus Ziegenbalg, appointed missionary of his most serene majesty the King of Denmark, to the East Indians and to their churches gathered in India, Halle in Saxony, Typeset and printed in the orphanage in MDCCXVI.

Chapter one deals with letters (pp.1-8) the Tamil alphabets, both the uyir (the vowels), and the mei (the consonants) and also the uyirmei (the mixed) are transliterated and the variations in each case is explained in Latin.

Chapter two explains pronunciation (pp.9-14). Ziegenbalg had used Latin letters to show how a few Tamil words were pronounced. In the case of words, he wrote in the first column the 25 Tamil words, then the transliterated form in the second column and their meanings in Latin, in the third column.

Chapter three dwells on nominative/substantive (pp. 15-29). In this chapter, the nominative, the genitive, the dative and the accusative forms of the Tamil nouns are explained, with Latin equivalents.[10] Chapter four deals with adjective (pp.30-44). In this chapter, Ziegenbalg explains the Tamil adjectives with their Latin equivalents and also he explains how each adjective is to be declinated.

Chapter five demonstrates pronoun (pp.44-59). In this chapter, we find how Tamil pronoun words are conjugated in the nominative, accusative, dative and genitive cases; and also here we find their Latin equivalents.

Chapter six dwells on verbs (pp.60-107). In this chapter, he explains how Tamil verbs are used in different numbers (singular and plural) and persons, i.e., first, second and third persons. He also outlines Tamil personal pronouns and how they are conjugated in singular and plural also in the first, second and the third persons along with their Latin equivalents.

Chapter seven dwells on particles (pp.108-117). In this chapter, Ziegenbalg explains how in Tamil language, prepositions could be attached to nouns and he also gives the Latin equivalents in each case. Likewise, we find also the adverbs, the interjections and the conjunctions.

Chapter eight is on syntax (pp.118-128). Ziegenbalg here shows the formation of sentences. For every formation of the sentence in Tamil, we have side by side, its Latin equivalents.[11]

Grammar was important for understanding and to speak Tamil. Among European languages the first printed grammar was 'Grammatica Damulica'. Tamil type fonts available in Europe have been proved by this, though Tamil fonts were made use in 1577 in India. A striking feature found missing in this book is the conventional dots atop the 18 Tamil consonants. Ziegenbalg used Greek and Latin grammar text models for writing Tamil grammar. He had mentioned colloquial language examples such as onnu, mannu, anju, venum etc.

Ziegenbalg had transcribed the sounds of alphabets and given 100 words for pronunciation in his work. We find three different models. At first we find words of colloquial Tamil given in colloquial Tamil (i.e., nannaka for nantraha; selai for silai). Some words of literary Tamil are given in colloquial language (i.e, penn for pon; veedu for wudu). Some colloquial terms are given in literary Tamil (i.e, pidavai for budawei: kezhvi for kelwi).

Fifty copies of the Tamil grammar printed in Latin had reached Ziegenbalg in London and the letter of Heinrich Julius Elers (16671728) dated 21 December 1715 also reached Ziegenbalg before sailing to Tranquebar. Walther who received a copy of Ziegenbalg's grammar commented and he opined that Ziegenbalg did not have enough time to elaborate the chapter on Tamil syntax. He also stated that the rules of Tamil syntax could have been better explained and beautifully arranged.[12] Thus we notice that Ziegenbalg's Grammatica Damulica of 1716 had been the first ever Tamil book printed in Europe. The book introduced Tamil to a wide circle of Europeans in Germany and England. It was reviewed in the journal *Acta Eruditorum* and it was warmly recommended to the readers in Europe. Thereafter several intellectuals in London and Berlin interacted.[13]

Tamil Grammar in Latin by Christoph Theodosius Walther: Printing at Tranquebar in South India in 1739

Christoph Theodosius Walther was born at Schildberg near Soldin in Brandenburg on 20 December 1699. He studied at Halle, was ordained and landed at Tranquebar on 19 June 1725, married in 1729, returned to Europe 1739, died at Dresden 29. April 1741. Walther who learnt Tamil wrote his Observationes Grammaticae. In his grammatical observations chapter one, he wrote on Tamil alphabets and letters in detail. Chapter two dealt with nouns and pronouns, chapter three on verbs, chapter four on syntax and finally we find an index.[14] Walther explained how certain Tamil letters like 'I' contract, how Tamil nouns changed according to their cases, how Tamil verbs were inflected according to their tenses and persons, and how sentences were constructed.

'Lectori Salutem' on the verso of the title page is dated, Tranquebar, 13 August 1739. Walther's knowledge of Tamil grammar and his application can be found. He mentioned that Tamil consonants and vowels are inseparable and said that '*uyir pannirendum udambu pathinettum*' i.e., Tamil vowels 12 in number are compared to life and consonants 18 are compared to the body. As a body without life and a life without a body cannot exist, so also in Tamil the vowels and consonants co-exist. In the chapter we find some Tamil words such as Udambu=body=corpus; Vasambu= Calamus acorus, Yeraalam=abundance=multitodo and Dharaalam= libera =confidential.

The printed Tamil grammar book of Beschi was bound along with the grammatical observations of Christoph Theodosius Walther the Protestant missionary which had highly displeased Beschi.[15] Nikolaus Dal (1690-1747) the fellow Protestant missionary described on 8 February 1740 how the Bishop of Mylapore rejoiced over Walther's appendix and welcomed it.[16] The diverse details included about measurements, terms of consanguinity, names of week days, months and years, naming the waning and with phases of the moon, zodiac signs and the life. Walther's observationes grammaticae had closely followed the study of Beschi's grammar.

The Spoken Tamil Grammar in Latin by Beschi: First Printing at Tranquebar in 1739, Second Printing at Madras in 1813 and the New Edition and Third Printing at Pondicherry in 1843

It may be said that the missionaries have shown a remarkable degree of individuality in learning Tamil and composing a grammar for the Portuguese, Dutch, German, French and Latin audience. Costanzo Giuseppe Beschi who tried his best to constantly improve the standard of his Tamil knowledge and sometimes adopted the best portion of the grammar composed by his immediate predecessors, applied his mind to write separately for colloquial Tamil and literary Tamil. The main reason he found was that the missionaries in the past were interested only in developing spoken Tamil grammar and neglected literary Tamil grammar. Beschi thus had moved to study literary forms and explained Tamil orthography, etymology and syntax in a very detailed manner. He also followed the Latin model as known to him and he did his best by widening the horizon of Tamil grammar.

Beschi's first grammatical work titled in Latin *Grammatica Latino-Tamulica ubi de Vulgari Tamulicae*, a grammar of common Tamil (titled Kodunthamizh in Tamil) was completed on 29 December 1728. Kodunthamizh had enabled the student to speak the language. Beschi's grammar of the lesser dialect therefore deals with all the traditional aspects of Tamil grammar. He wrote on ezhuthu (orthography), sol (etymology and syntax), porul (interior and exterior matter), yappu (prosody), and ani (rhetoric).

Beschi although does not make mention of the Tamil grammar in Portuguese by Balthasar da Costa in his grammar, we may conclude that he must have extensively used it. We find in his sixth chapter dealing with syntax and irregular verbs, whatever is found in the work of Balthasar da Costa. The information provided by Beschi includes some suffixes such as, um, o, a, aiyo, anro, yeen, aakkum, mun, pin, muthi, padi, appuram, appaal, mel, kizh, maathiram, alavu, thanai, varaikkum, mattum, athikam, paarka, kaatilum, udanay, illaamal, allaamal, anri, inri, aanalum, enkil, aam, onril, pola, idaividaamal, illaavittal, ozhiya, thavira, oruppada, orumikka, yerakuraiya, ara, azha, kitta, vilaga, kaduga, muduga and mai. Even some examples of sentences cited by Balthasar da Costa such as '*Ni visuvaskiraye, appothu avan visuvasithan*', '*Ni visuvasikiraye appothu nan visuvasipaen*', '*Ni visuvaskiraye athukku munay nan visuvasipaen*' had been mentioned by Beschi.[17]

Here he says that some other author (without naming Balthasar da Costa) had mentioned such usages, but they were not found common among Tamils of his times. Thus we find that Balthasar da Costa's Portuguese-Tamil grammar had been used by Beschi like Ziegenbalg who also used the same while composing Tamil grammar in Latin. But Beschi did not indulge in plagiarism and copied extensively and verbatim like Ziegenbalg from the chapters of Balthasar da Costa on adjectives, pronouns, and verbs. Beschi had followed to a larger extent the dictionary model of Filippo Venuti (1531-1587) the Italian whose book was printed in 1582 titled '*Dittionario volgare et latino nel quale si contiene come i vocaboli italiani si possono dire et esprimere latinamente per M. Filippo Venuti da Cortona con un dittionario delle voce latine similie à quello del Calepino co' numeri per ritrouar nel volgare la loro significatione aggiunto dal medesimo in questa nuova impressione*, In Venetia: appresso gli heredi di Luigi Valuassori & Gio. Domenico Micheli.

Beschi in his in preface to grammar of common Tamil had cleared up a major source of confusion by distinguishing short and long forms of two vowels (o and e) both as independent vowels and then joined with consonants. Earlier grammars had prescribed the use of a line over the vowel to distinguish the short form from the long form but this method had fallen into disuse. Beschi had rationalized the orthography for the three separate forms of r (r, ra, RA). The changes introduced by Beschi in the Tamil script are significant to note. The Tamil alphabet always had the long and the short e and o, though the written form did not have separate letters. According to Tholkappiyam and Nannul the distinction between these vowels was that the short e and o carried a pulli or dot above the letter. Beschi changed the system slightly. To the short letter ye (e) he added a slanting stroke and wrote the long form as yee (ee). Similarly the short 'o' he added a loop at the 'ye' (e) is represented by a kompu and Beschi added a loop at the top to this kompu to make it represent the long sound. These changes an important contribution by Beschi had become permanent features of the Tamil script.

Beschi's grammar of the common dialect of the Tamil language dealt with on the special use of certain words. According to him in Tamil several nouns which must not be used promiscuously denoted the infancy of men, or of animals or of trees. Thus infants and children are called kulandai and pillai; of which the last is also used of a young coconut palm, thennam pillai; a squirrel anil pillai; a mongoose keeri pillai and a parrot kili pillai. The term kunju is used of birds, of mice, of fishes and of insects thus pura kunju, a young pigeon; kozhi kunju a chicken; yeli ķunju mice; meen kunju small fish etc. The common word for the young of four-footed animals is kutti. Thus, singa kutti, kudirai kutti, naay kutti, punai kutti etc is used. It is also used of serpents; as pambu kutti. The calves of cows and buffaloes cannot be called simply kutti but kanru or kanru kutti; thus pasuvin kanru or pasuvin kanrukutti; yerumai kanru or yerumai kanru kutti. Foals of horses or asses are either kutti or mari. Thus kudirai mari, kazhuthai mari had been used. Of harts, camels and elephants, we may say either kutti or kanru; maan kanru, ottai kanru and yaanai kanru. All young trees, whilst they are little (the palm tree excepted) are kanru; maang kanru, veppang kanru, puliyang kanru, murungai kanru etc. Naatru is a young plant or seedling.

Beschi stated that Aaan means man; penn means woman. Pillai is often added but thus used pillai does not always mean as child; but often person, young or old. Aan pillai-a male; penn pillai a female was used. Aaan and penn may be used of infants, animals and birds; aan kulandai–a male infant; aan maan–a male deer, a stag; aan pantry–a boar, hog; penn kudirai a mare.

Different terms are used to distinguish the male and female of cattle at various stages in their growth. A small calf is kaazhai kanru; a small female calf is kidaari kanru; later the male calf is kaazhai, later the female calf is kidaari; when fully grown the ox is yeruthu and when fully grown cow is pasu, the male buffalo at every age is kida or kada, but when young it may be kida kanru. The female buffalo is always yerumai and if young it may be yerumai kanru. Sometimes the male of dogs, cats, jackals, monkeys etc is kaduvan. A male monkey is called kaduvan kurangu.

Rarely the female is pettai as pettai punai, a she-cat but generally penn is used. Many special words are used as kalai a stag, kidai or kada a male sheep or goat. The male of birds is saval or seval; the female pettai and these two words may be placed before or after the specific name of every bird. Thus saavarkozhi or kozhiseval, a cock; pettai kozhi or kozhi pettai, a hen; saavar pura, a male pigeon pettai pura, a female pigeon etc had been used.

In the introduction to his grammar of the literary dialect Beschi explained that he has omitted porul and ani because rhetoric would be well known to European students and ezhuthu had been exhaustively treated in his grammar of Kodunthamizh. The work is thus divided into only two parts; the first part dealt with letters and words and the second part, titled 'De Poesi Tamulica', with prosody including, at the same time, a few notes on Tamil poetry. Since Beschi had compared the structure of European language grammars and pointed out the similarities and differences in Kodunthamizh, T.P. Meenakshi Sundaram opined that the work was a 'transfer grammar'.[18] This view needs to be reconsidered for Beschi emphasized and pointed out the fine use in Tamil. Speaking of 'the use of the infinitive' in section five of Kodunthamizh Beschi made some remarks and it is important to mention here. He stated 'They (the Tamils) use the infinitive elegantly, as the Latins by the figure of ellipsis, when the verb incipio, to begin, & c., being omitted, they elegantly fill the sentence, by a crowded series of infinitives: eg. extensae apparere nubes, undique obscurari caelum, fulgura micare, mugire tonitura & c'. Beschi who began to write grammar of common Tamil had completed the work on 29 December 1729.[19] The Tranquebar missionaries desired to publish his work of the grammar of common Tamil and it was printed at Tranquebar press in 1739.[20] Dom Jose Pinheiro the eighth Bishop of Mylapore (consecrated on 24 March 1726-died on 15 March 1744) added a preface dated 2 November 1737 and he blessed the book.[21] This work of Beschi is the first bilingual Tamil grammar printed at Tranquebar in India.[22] Though Ziegenbalg's Tamil grammar in Latin was earlier printed in Halle in 1716, the grammar of common Tamil by Beschi is very important because he had set forth a scheme of reforming the Tamil scripts. Beschi's Kodunthmizh had attracted much attention and we find that it was translated by Christopher Henry Horst and got printed in Madras in 1806.[23] Again Francis Whyte Ellis had printed the Kodunthamizh at the Fort St. George College Press in Madras in 1813 for the second time.[24] After 30 years, Louis Savinien Dupuis in Pondicherry had printed Kodunthamzih for the third time in 1843. This is a new edition and we find foot notes prepared and given in some places. In the printed title of the book we find mention of the expression '*Nova Editio cum notis et Compendio grammaticae de Elegantiori dialect Senthamizh Dicta*'.[25] The book contains viii+215+28 pages.

The Tamil Grammar in English by Johann Philipp Fabricius: Printing at Madras in 1789

Johann Philipp Fabricius was born at Kleeberg near Frankfort on 22 January 1711. He studied at Giessen and Halle, was ordained at Copenhagen

1739, embarked at Gravesend on 1 April 1740 and landed at Cuddalore on 28 August 1740. He went to Tranquebar and stayed there and later he moved over to Madras in November 1742 when he became the successor of Benjamin Schultze. He stayed in Madras and died there on 23 January 1791. Fabricius learnt Tamil, and began to write Tamil grammar in English. He published the work titled 'A grammar for learning the principles of the Malabar language properly called Tamul or the Tamulian language by the English missionaries of Madras', printed at Wepery near Madras, in the year 1778. The book has 63 pages.[26] It got reprinted in 1789 with the title 'A Grammar for learning the principles of the Malabar language, properly called Tamul or the Tamulian language'. The names of Fabricius and Breithaupt are found within the Tamil imprint. The book has 63 pages.[27] There are nineteen chapters such as of the letters, syllables and pronunciation, of nouns substantive, of nouns adjective, of the observations with regard to the use of cases of the nouns and pronouns, conjugation of verbs, of verbs neuter and verbs passive and of forming such verb active as express the getting a thing done, of the conjugations of verbs in the negative way, of the way how to express the prerter-perfect, and preter-pluperfect and the subjunctive, of such verbal nouns as are constructed like verbs, of defective verbs, of conjoining verbs together, some further observations with regard to the verbs, of the construction, of some particles, of some usual termination of words that signify persons, of the numbers, of the day, week, months and years in Tamil (from 1747 to 1806) and a collection of Tamil proverbs.

In the introduction Fabricius stated thus: 'the high, abstruse and poetical Tamul, quite different from the ordinary and common, that is used by all the people, we have nothing to do with, and leave it to those that are fond of it'.[28] Fabricius mentioned that 'in this grammar, one finds a peculiar collection of proverbs, written in somewhat odd English and translated from Tamil: if a flood runs over a dam, will it return through one cries ever so much? Will the shame contracted for half a *kasu* (cash) leave one by giving a thousand? The child that cries will get the milk. When grain is scarce, everything is scarce'.[29] The European reader untrained in this kind of proverbial thinking would probably have been puzzled to read these lines. The grammar for learning the principles of the Malabar language by Fabricius was reviewed in the journal of '*Allgemeine Literaturze*'.[30] Thus we notice that Tamil grammar was made available in English for wider audience.

To conclude it may be said that as the Europeans intended to keep their knowledge within their communities, there was initially no necessity in publishing all manuscripts prepared on Tamil grammar. Many travelled a long road before they saw the light of printing. Hence we need to see the originals composed and the printed versions and we find surprisingly very important omissions as in the case of Antão Proença's grammar and also omission of paragraphs in the printed dictionary where he spoke on the aspects of Tamil grammar.

Early missionary Tamil grammars look very similar and the presentation had been from the practical orientation. They had been intended to serve as textbooks. However, the Tamil grammar of Beschi the missionary in many ways had differed strikingly from others and it was an improvement. The approach is evident in his description of data which he referred to as found in the native grammar of nannul. Beschi also had tried to fit well his Tamil grammar into the Latin grammar framework. Thus he sheds new light on the complex nature of the phenomenon which he confronted with and we find his methodological challenges besides encountering the difficulties he faced in the regional variations of Tamil as a fieldworker.

Notes

1. Daniel Jeyaraj, *Tamil Language for Europeans: Ziegenbalg's Grammatica Damulica (1716): Translated from Latin and Tamil*, Wiesbaden, 2010, p. 16.
2. Ibid., pp. 19-20.
3. School of Oriental and African Studies Library, London (hereafter SOASL), B. Ziegenbalg, *Grammatica Damulica, quæ per varia paradigmata, regulas & necessarium vocabulorum apparatum, viam brevissimam monstrat, qua Lingua Damulica seu Malabarica, quæ inter Indos Orientales in usu est, & hucusque in Europa incognita fuit, facile disci possit: in Usum eorum qui hoc tempore gentes illas ab idolatria ad cultum veri Dei, salutemque æternam Evangelio Christi perducere cupiunt: In itinere Europao, seu in nave Danica, concinnate a Bartholomaeus Ziegenbalg*, Halae Saxonum: Litteris & impensis Orphanotrophei, MDCCXVI (hereafter Grammatica Damulica), pp. 4559.
4. Ibid., pp. 61-107.
5. Daniel Jeyaraj, Tamil Language for Europeans, p. 22.
6. Archiv der Franckesche Stiftungen (hereafter AFSt), Halle, M1 C7:2, fl.1, Arno Lehmann, ed., *Alte Briefe aus Indien: Unveröffentlichte Briefe von Bartholomäus Ziegenbalg 1706–1719*, Berlin, 1957, p. 433; Daniel Jeyaraj, Tamil Language for Europeans, pp. 22-3.
7. Daniel Jeyaraj, Tamil Language for Europeans, p. 23.
8. Arno Lehmann, Alte Briefe aus Indien, p. 436.
9. Sudasien Institut, Universitat Heidelberg, nsp50.12A80/153 rara. Grammatica Damulica, pp. xiv and 128; A facsimile reprint was done by Burchand Brentjes & Karl Grallus, *Grammatica Damulica von Bartolomaeus Ziegenbalg*, Halle, 1985 and Halle, 2005.

10. Ibid., pp. 8-26.
11. Ibid.
12. AFSt, M1 B2:26; HB, vol. II, continuation 21, p. 724.
13. B. Ziegenbalg, 'Grammatica Damulica', *Acta Eruditorum*, 1717, pp. 253–5.
14. Cecil H. Green Library, Standford University, California, *Christophoro Theodosio Walthero, Observationes Grammaticae quibus Linguae Tamulicae Idioma Vulgare inusum operariorum in messe Domini intergentes Malabares Dictas illustratur*, Trangambariae, MDCCXXXIX, The book has 60 pages. Chapter I: Tamil Letters (pp. 315), Chapter II: Nouns and Pronouns (pp. 15-21), Chapter III: Verbs (pp. 22-38), Chapter IV: Syntax (pp. 38-58), Index (pp. 59-60). On verso of the title Lectori Salutem we find mention of Tranquebar 13 August 1739. We also find a copy of the same book at the Archives of the United Theological College Library and Archives, Bengaluru; Universiteits Bibliotheek Gent, Belgium; Library of the School of Oriental and African Studies, London (vide EB.73.15/2) and Bodleian Library, Oxford (vide Drave.e.68).
15. Gregory James, *Col Porul: A History of Tamil Dictionaries*, Chennai, 2000, p.154.
16. AFSt, M1 B26: 29.
17. Bayerische Staatsbibliothek, Munich, *Grammatica Latino-Tamulica, ubi de Vulgari Tamulicae Linguae Idiomate, Kodunthamizh dicto, ad usum Missionariorum Soc. IESU, Auctore P. Constantino Iosepho Beschio, Ejusdem Societ. In regno madurensi Missionario AD MDCCXXVIII*, Trangambariae, Typis Missionis Danicae MDCCXXXIX.
18. T.P. Meenakshi Sundaram, 'Tamil Letters and Sounds of the Seventeenth Century', *Collected Papers of Prof. T. P. M, Sixty-first Birthday Commemoration Volume*, Annamalai Nagar, 1961, pp. 225-31.
19. British Library, London (hereafter BL), OR. MS 13586. This manuscript has a printed slip that reads, 'written in a beautiful hand by Constantius Iosephus Beschi, missionary in Madurai 29 December 1729'. There is also another manuscript. See BL, Oriental and India Office Collection (hereafter OIOC), OR. MS 13044; See, *Grammatica Latino-Tamulica in Qua De Vulgari Tamulicae Linguae Idiomate Kodun Thamzih Dicto, Fusisus Tractatur, Auctore P. Constantio-Josepho Beschio, Societatis Jesu, In Regione Madurensi, Apud Indos orientales, Missionario, NOVA EDITIO Cum notis et Compendio Grammaticae De Elegantiori Dialecto Sen Thamizh Dicata, An uno Missionario Apostolico Congragationis Missionum Ad Exteros*, PUDICHERII, E Typographio Missionariorum Apostolicorum Dictae Congregationis, 1843. There are five chapters, such as (1) De Litteris (letters), (2) De Nomine (nouns), (3) De verbo (verbs), (4) De Syntaxi (syntax) and (5) De Variis quotidiano usui necessaris (everyday usages). This book contains 216 pages. The pages are numbered from 1 to 200 and had 200 paragraphs. In the beginning the preface of the editor is given. Editoris Prefatio, J.M.J., Omnibis Operariis Apostolicis, in Tamulicis Regionibus Degentibus, Salutem in domino plurimam. The preface has four pages and at the end Pudicherii, 6 Kal.Julii 1843 is recorded. After this the preface of the author is mentioned: Auctrois Prefatio: Religiosis Linguae Tamulicae candidates Constantius Joseph Beschius Salutem in Domino plurimam.

The preface runs for four pages and at the end E Missione Madurensi.4.Kal. Jan. 1728 is recorded citing the date when Beschi completed the work. See also, *A Grammar of the Common Dialect of the Tamul Language called Kodunthamizh; Composed for the use of the missionaries of the Society of Jesus [...], translated from the original Latin by George William Mahon*, Vepery/Madras: Printed by Reuban Twigg at the Christian Knowledge Society's Press, 1848.

20. SOASL, EB.73.20; see another copy in EB 73.15/1.
21. Det Kongelige Bibliotek (hereafter KB), Kobenhavn, 47-525.
22. AFSt, M 1H 1B:12.
23. SOASL, *A Grammar of the Common Dialect of the Tamulian Language Called Kodunthamizh Composed by R.F. Const. Joseph Beschi, Jesuit Missionary after a Study and Practice of Thirty Years, Translated by Christopher Henry Horst*, 1806, Madras, Printed at Vepery. The book has pages vii+192, with four folding charts.
24. *Grammatica Latino-Tamulica, in qua de Vulgari Tamulicae Linguae Idiomate, Kodunthamizh dicto, fusius tractatur, Auctore P. Constantino Iosepho Beschio, E Societate IESU, et in Regione Madurensi, apud Indes Orientale, Missionario*, Apud Madraspatnam, E Typographeo Collegii, 1813. The next page runs as follows: *R.P. Constanti Josephi Beschii vulgaris Linguae Tamulicae Indiomatis Grammaticae nova heac Editio, primae Editionis Exemplari, cum MS. Eoden fere tempore exarato, diligenter collato, et Paradigmatibus quibusadm exaucta, studiosrum in usum, nunc tandem in lucem proditi, cura atque opera Collegi Madraspatensi Praepositorum*. F.W. ELLIS, E.C GREENWAY, W.WAYTE, J.MOUSLEY, W.OLIVER, J.BABINGTON, J.MACKERRELL, A.D.CAMPBELL, COLLEGI MADRASPATENSIS PRAEPOSITI. The next page runs as follows: PRIMAE EDITIONIS PRAEFATIO. RELIGIOSIS LINGUAE TAMULICAE CANDIDATIS CONSTANTIUS JOSEPH BESCHIUS, salutem in Domino plurimam. At the end of the preface, we find E Missione Madurensi. 4.Kal. Jan. 1728 is recorded.
25. Bibliothèque Asiatique Missions Etrangères de Paris (hereafter BAMEP), B 2305, *Grammatica Latino-Tamulica, in qua de Vulgari Tamulicae Linguae Idiomate, Kodunthamizh dicto, fusius tractatur, Auctore P. Constantino Iosepho Beschio, E Societate IESU, in Regione Madurensi, apud Indos Orientales Missionario NOVA EDITIO CUM NOTIS, ET COMPENDIO GRAMMATICAE de elegantori Dialecto Senthamizh dicta, ab uno Missionario Apostolico, Congregationis Missionum Ad exteros*, Puducherii, E typographio Missionariorum Apostolicorum dictae Congregations, 1843.
26. Bodleian Library, Oxford, Drave.e.80, Johann Philipp Fabricius, *A Grammar for Learning the Principles of the Malabar Language, Properly Called Tamul or the Tamulian Language, by the English Missionaries of Madras*, Printed at Wepery near Madras in the year 1778.
27. KB, 167-1665, Johann Philipp Fabricius, *A Grammar for Learning the Principles of the Malabar Language, Properly Called Tamul or the Tamulian Language, by the English Missionaries of Madras,* Printed at Vepery near Madras, the second edition in the year of Our Lord, 1789. Copy of the same book is available at Library of the School of Oriental and African Studies, London (vide EB.78.27), Bodleian Library, (vide Drave.e.90) and BL, London (1609/4372).

28. Ibid., p. 3.
29. Ibid., p. 58.
30. Johann Philipp Fabricius, A Grammar for Learning the Principles of the Malabar Language, Properly Called Tamul in '*Allgemeine Literaturze*', vol. IV, no. 378, 1794.

GRAMMATICA
DAMULICA,
Quæ
per varia paradigmata, regulas & necessarium vocabulorum apparatum,
VIAM BREVISSIMAM
monstrat,
Qua
LINGUA DAMULICA
Seu MALABARICA, quæ inter Indos Orientales in usu est, & hucusque in Europa incognita fuit, facile disci possit;
in
Usum eorum
Qui hoc tempore gentes illas ab idololatria ad cultum veri Dei, salutemque æternam Evangelio Christi perducere cupiunt:
In itinere Europæo, seu in nave Danica, continuata
a
BARTHOLOMÆO ZIEGENBALG,
Serenissimi Regis Daniæ Missionario inter Indos Orientales, & ecclesiæ ex Indis collectæ Præposito.

4.1. The title page of Grammatica Damulica of Bartholomäus Ziegenbalg printed in Halle in 1716 (Courtesy: Bibliothek der Franckesche Stiftungen, Halle, Miss: C148).

CAP. II.
de
LECTIONE AC PRONUNCIATIONE HUJUS LINGUÆ.

Hæc lingua legitur a sinistra ad dextram, sicut omnes linguæ Europææ. Sed non nisi ab iis, qui eam juxta verborum copiam probe intelligunt, recte legi potest.

In scriptione hujus linguæ neque commata, neque puncta, neque cola vel semicola, neque signa interrogationis vel exclamationis, neque ulla alia distinctionis signa usurpantur; & tamen per se juxta structuram intellectu est facilis, clara & perspicua iis, qui verborum copia instructi sunt ac ejus indolem sibi familiarem reddiderunt.

Primo legi ac pronunciari possunt simplices voces sine interjectis literis finalibus hunc in modum:

பிதா	*Pita*	Pater
மாதா	*Mata*	Mater
கிருபை	*Kirubei*	Gratia
ஆதி	*Adi*	Principium
நீதி	*Nidi*	Justicia
ராசா	*Rascha*	Rex
உதவி	*Udawi*	Auxilium
மகிமை	*Magimei*	Gloria
பிசாசு	*Pischaschu*	Diabolus
பூமி	*Pumi*	Terra
அறிவு	*Ariwu*	Notitia
நினைவு	*Ninawu*	Cogitatio
மனசு	*Manaschu*	Voluntas
புகை	*Pugei*	Fumus
சாதி	*Schadi*	Genus
கனி	*Kani*	Fructus

B

4.2. On the reading and pronunciation of Tamil language by Bartholo-mäus Ziegenbalg's Grammatica Damulica printed in Halle in 1716, page 9 (Courtesy: School of Oriental and African Studies Library, London University).

A. M. D. G.
GRAMMATICA
LATINO-TAMULICA,
Ubi de
Vulgari Tamulicæ Linguæ
Idiomate
கொடுந்தமிழ்
dicto,
ad Usum
MISSIONARIORUM
Soc. IESU.
AUCTORE
P. CONSTANTIO IOSEPHO BESCHIO,
Ejusdem Societ. in Regno Madurensi
MISSIONARIO.
A. D. MDCCXXVIII.

TRANGAMBARIAE,
Typis Missionis Danicæ, CIƆ IƆ CCXXXIIX.

4.3. The title page of Spoken Tamil Grammar in Latin by Costanzo Giuseppe Beschi Printed in Tranquebar Press in 1739 (Courtesy: Bayerische Staatsbibliothek, Munich).

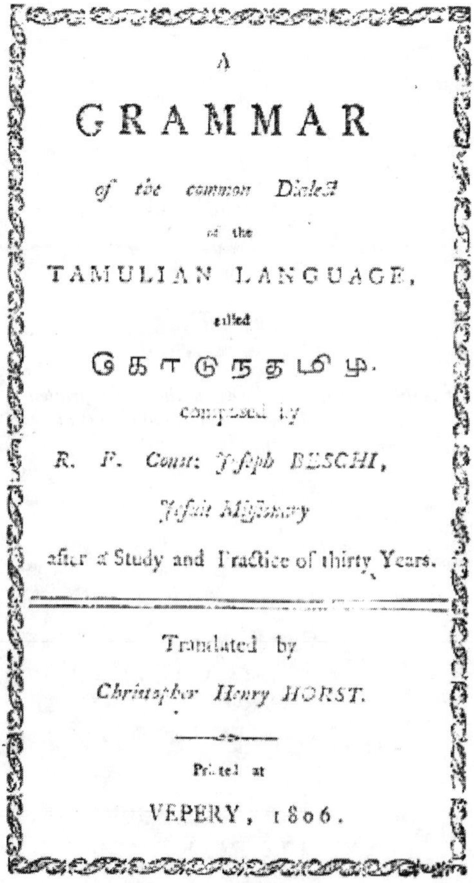

A

GRAMMAR

of the common Dialect

of the

TAMULIAN LANGUAGE,

called

கொடுந்தமிழ்.

composed by

R. F. Const: Joseph BESCHI,

Jesuit Missionary

after a Study and Practice of thirty Years.

Translated by

Christopher Henry HORST.

Printed at

VEPERY, 1806.

4.4. The title page of grammar of the Kodunthamizh composed by Beschi, translated by Christopher Henry Horst and printed in Madras in 1806 (Courtesy: School of Oriental and African Studies Library, London University).

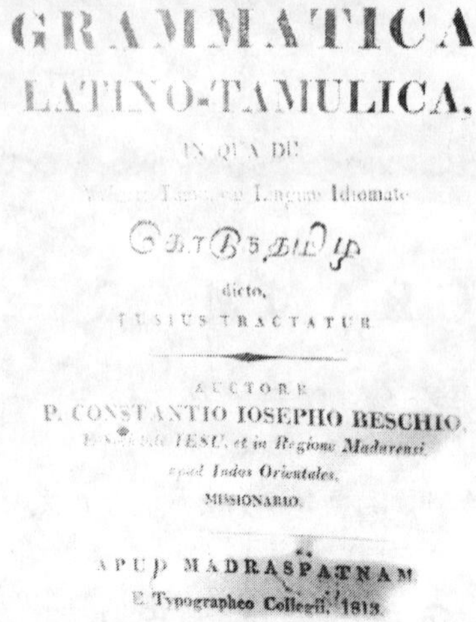

GRAMMATICA
LATINO-TAMULICA,
IN QUA DE
Vulgari Tamulicæ Linguæ Idiomate
கொடுந்தமிழ்
dicto,
FUSIUS TRACTATUR

AUCTORE
P. CONSTANTIO IOSEPHO BESCHIO,
Societate IESU, et in Regione Madurensi
apud Indos Orientales,
MISSIONARIO.

APUD MADRASPATNAM
E Typographeo Collegii, 1813.

4.5. The title page of Beschi's Kodunthamizh reprinted at Fort St. George College press in Madras in 1813 (Courtesy: Author's private collection of rare books)

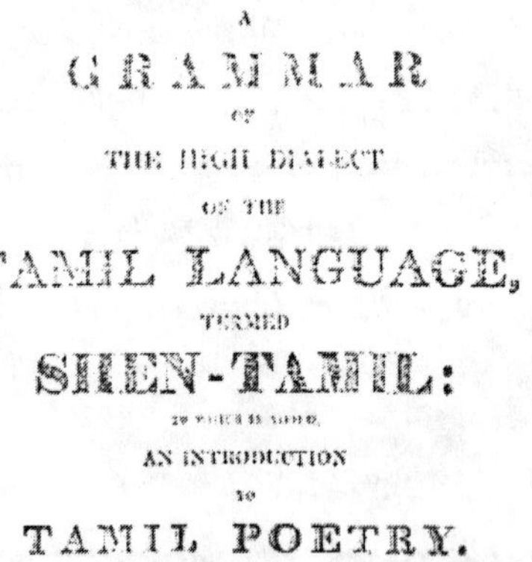

A
GRAMMAR
OF
THE HIGH DIALECT
OF THE
TAMIL LANGUAGE,
TERMED
SHEN-TAMIL:
TO WHICH IS ADDED,
AN INTRODUCTION
TO
TAMIL POETRY.

BY THE REVEREND FATHER C. J. BESCHI,
Jesuit Missionary in the Kingdom of Madura.

TRANSLATED FROM THE ORIGINAL LATIN,
BY
BENJAMIN GUY BABINGTON,
Of the Madras Civil Service.

Madras:
PRINTED AT THE COLLEGE PRESS, 1822.

4.6. The title page of Shenthamizh translated by Benjamin Guy Babington printed at the College press in Madras in 1822 (Courtesy: Church Mission Society Library and Archives, Birmingham).

A

GRAMMAR

For learning the Principles of the

MALABAR LANGUAGE,

Properly called

TAMUL or the TAMULIAN

LANGUAGE

By the English Missionaries of Madras.

Printed at Vepery near

MADRAS

The second Edition

in the Year of our Lord 1789

4.7. The title page of the second edition of Tamil grammar in English by Johann Philipp Fabricius printed at Vepery in Madras in 1789 (Courtesy: Bodleian Library, Oxford).

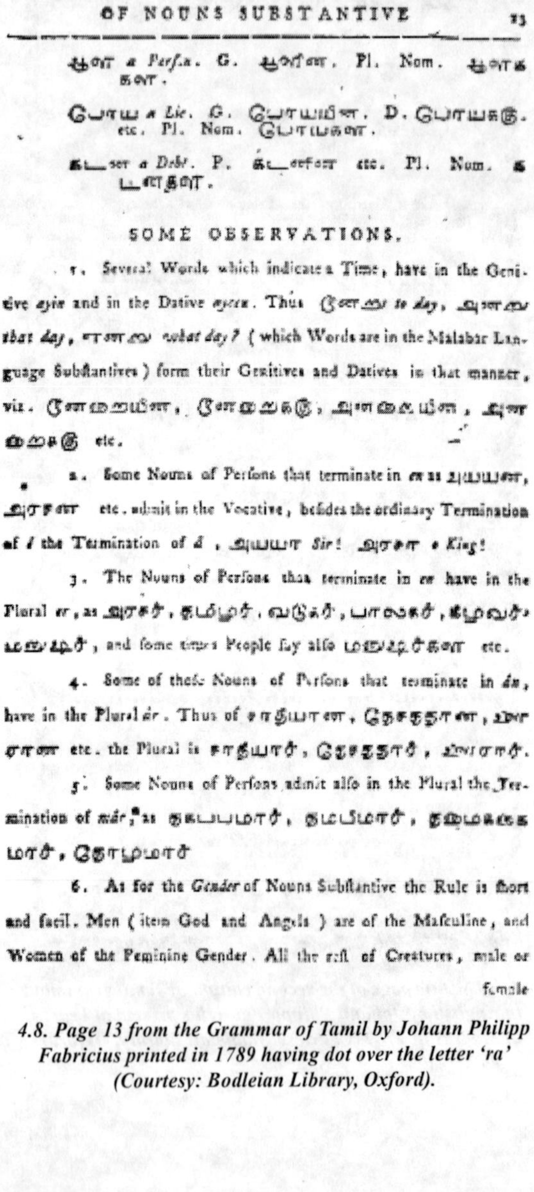

OF NOUNS SUBSTANTIVE 13

ஆள் a Person. G. ஆளின். Pl. Nom. ஆளக்கள்.

பொய் a Lie. G. பொய்யின். D. பொய்க்கு. etc. Pl. Nom. பொய்கள்.

கடன் a Debt. P. கடனின் etc. Pl. Nom. கடன்கள்.

SOME OBSERVATIONS.

1. Several Words which indicate a Time, have in the Genitive *ayin* and in the Dative *aycku*. Thus இன்று *to day*, அன்று *that day*, என்று *what day?* (which Words are in the Malabar Language Substantives) form their Genitives and Datives in that manner, viz. இன்றையின், இன்றைக்கு, அன்றையின், அன்றைக்கு etc.

2. Some Nouns of Persons that terminate in *en* as ஐயன், அரசன் etc. admit in the Vocative, besides the ordinary Termination of *é* the Termination of *á*, ஐயா *Sir!* அரசா *a King!*

3. The Nouns of Persons that terminate in *en* have in the Plural *er*, as அரசர், தமிழர், வடுகர், பாலகர், கிறிஸ்தவர், மனுஷர், and some times People say also மனுஷர்கள் etc.

4. Some of these Nouns of Persons that terminate in *án*, have in the Plural *ár*. Thus of சாதியான், சேசுநாதன், ஸ்ரீராமன் etc. the Plural is சாதியார், சேசுநாதர், ஸ்ரீராமர்.

5. Some Nouns of Persons admit also in the Plural the Termination of *már*; as தகப்பமார், தம்பிமார், தமையக்கமார், தோழமார்

6. As for the *Gender* of Nouns Substantive the Rule is short and facil. Men (item God and Angels) are of the Masculine, and Women of the Feminine Gender. All the rest of Creatures, male or female

4.8. Page 13 from the Grammar of Tamil by Johann Philipp Fabricius printed in 1789 having dot over the letter 'ra' (Courtesy: Bodleian Library, Oxford).

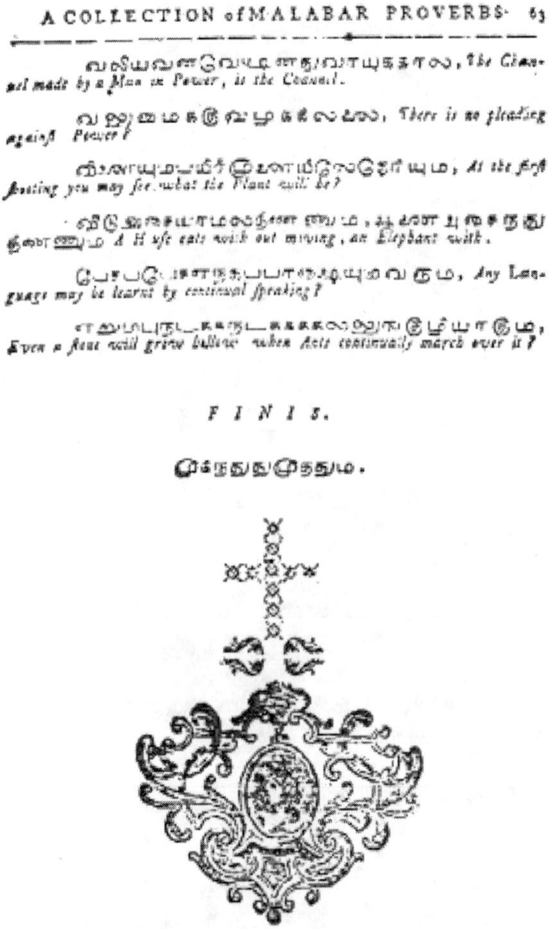

A COLLECTION of MALABAR PROVERBS. 63

[illegible], *The Channel made by a Man in Power, is the Counsel.*

[illegible], *There is no pleading against Power?*

[illegible], *At the first sprouting you may see what the Plant will be?*

[illegible], *A Horse eats without moving, an Elephant with.*

[illegible], *Any Language may be learnt by continual speaking?*

[illegible], *Even a Stone will grow hollow when Ants continually march over it?*

F I N I S.

[illegible].

4.9. The last page from the Tamil grammar in English by Johann Philipp Fabricius printed in 1789 containing Tamil proverbs (Courtesy: Bodleian Library, Oxford).

5

Louis Savinien Dupuis in Pondicherry and Printing Tamil Grammar in French, Tamil and Latin and its Reception around the World, 1845-63

Louis Savinien Dupuis the MEP missionary in Pondicherry was greatly inspired by the Tamil grammatical writings of Costanzo Giuseppe Beschi. Therefore he planned and decided to make some attempts to study Tamil grammar which could be highly successful. Louis-Marie Mousset another MEP missionary who came to Pondicherry and served in the mission station like Kamanayakkanpatti where Beschi served was highly influenced by knowing and reading the writings of the famous Jesuit. It has been planned by these two MEP missionaries in Pondicherry in a joint venture to compose grammatical works for the use of the people of various walks of life, like the natives in Pondicherry, the local Tamil seminarians, European residents in the French colony of Pondicherry and the coolie labour Tamil migrants to Bourbon and Martinique islands. Dupuis was the then pro-vicar Apostolic of the Pondicherry Vicariate and Mousset the then professor at the seminary in Pondicherry felt the urgent need to compose multiple grammars and the necessity for preparing these grammar works. Thereafter they made efforts, wrote and published several works. In this chapter we shall see how Dupuis had printed Piranchu Ilakkana Noolsurukkam in 1845, Ilakkana Surukkam in 1845, Lathin Ilakkana Noorsurukkam in 1845 at the Catholic Press in Pondicherry. We shall also examine how he later published Ilakkana Noolaatharam in 1849 and Grammaire Française-Tamoule in 1863.

Piranchu Ilakkana Noolsurukkam: The Elements of French Grammar in Tamil Printed in 1845

Dupuis in his letter dated 16 March 1844 to Jean Tesson had mentioned that the work of French-Tamil grammar was in progress.[1] Mousset vide his letter dated 19 April 1845 to Jean Tesson had said that the work of French-Tamil grammar had been successfully completed.[2]

Piranchu Ilakkana Noolsurukkam the book appeared in print in 1845 was composed by Louis Savinien Dupuis and Louis-Marie Mousset. The title page of the book does not directly refer to the authors but mentions (in plural) the priests of the société des missions étrangères de Paris. In the missionary correspondence of letters it has been mentioned that they were the authors. The total pages of the printed book are 170. The contents of the book had been given at the end.[3] The errata of two lines had only been provided. Thus the book had been printed very carefully without mistakes. It is gleaned that the book was prepared by the MEP missionaries for the purpose of learning French through Tamil. It was printed with the permission of the MEP superior.

In the book we find the French alphabets/letters that are twenty five in number which had been given vide page 1. The proverbs in French are given vide pp. 11-13. Catholic prayers in French had been furnished (Pater Noster, Ave Maria, etc) vide pp. 14-16. The general details of yezhuththuu athigaaram (on scripts/letters) are given vide pp. 26-27. Then we find some stories in French printed of eminent people like Emperor Theodose the Great, Emperor Constantine, Williams-Bealde and Francis Xavier who had led exemplary lives on this earth.

The book thereafter actually begins with an introduction and deals with solliyial (words) and Dupuis explains the ten types of words (sol) such as the noun or the substantive, the article in chapter two, the adjective in chapter three, the pronoun in chapter four, the verb in chapter five with examples of present tense, past tense and future tense in chapter five, the participle in chapter six, the preposition in chapter seven, the adverb in chapter eight, the conjunction in chapter nine and the interjection in chapter ten. Dupuis proceeds further on syntax in part two of the book and he explains the accent signs used in French.[4] A chapter is devoted for the style and method translation with example of passages in French and Tamil.[5]

In fine, the book is composed for learning French through Tamil grammar pattern. Hence it may be called a functional grammar book. Further the French grammar has been well compared by Dupuis and presented which is a remarkable feature found noticed. The book had been aimed at and prepared mainly for the use of the seminarians in Pondicherry. The French grammar in brief is written thus for the Tamil audience. This book on the elements of French grammar in Tamil has been composed by Dupuis and Mousset chiefly based on the pattern of linguistics where the rules have been classified and compiled on taxonomic theory. This is the first book written in Tamil language on French grammar and also got printed for the first time.

Lathin Ilakkana Noorsurukkam: The Elements of Latin Grammar in Tamil Printed in 1845

Dupuis in his letter dated 16 March 1844 to Jean Tesson had mentioned that the work of the Latin-Tamil grammar was in progress.[6] Mousset vide his letter dated 19 April 1845 to Jean Tesson had said that the Latin-Tamil grammar had been successfully finished. He also mentioned that they were authors of the work.[7]

Lathin Ilakkana Noorsurukkam was printed at the Catholic press in Pondicherry in 1845.[8] It is mentioned in the title page of the book that it was prepared by the MEP priests who were nominated/ appointed and the book was printed by the order of the superiors. It is gleaned from the missionary correspondence of the letters that Dupuis and Mousset were that authors of the book. The details of the contents had been printed at the end.[9] There is the erratum of three words found at page 227. Hence it is known that the book had been printed very carefully without many mistakes.

The printed book contains three parts. Part one deals with solliyal (words) and part two deals with thodar mozhi iyal (linguistics) and part three deals with translation of verses from Latin into Tamils and the differences between the two languages.

The book has an introduction in Tamil. It states that Latin language had assumed great significance and priority over other languages around the world because all the church doctrine and divine worship had been composed in that language. The greatness of Latin language was that the God's word and the Holy Scriptures had been available in that language. All the commentaries of the Holy Scriptures, sermons and teachings of the Church had been also found in the various treatises in Latin. Hence from time immemorial, all the nations, all the scholars, the elders and the educated have learnt Latin. Therefore in this Tamil land too, in order to make people understand all the church doctrine and other teachings of the church Latin language should be made known to the Tamils. A Latin grammar was therefore found very necessary to serve as a foundation since the available Latin grammar was inadequate. Hence it had had been decided by them to compose and print it.

It has been also mentioned that learning Latin was commonly accepted in France through the printed books. But the missionaries have decided not to translate such books for use in Pondicherry since there is vast difference between Tamil and French language. Hence the significant and the related Latin grammar items were composed and whatever feasible had been done very fresh in the best possible way. It has been also said that they have added newly the matter related to translation of Latin into Tamil and words and the verses have been shown explaining the ways and means and the methods for translation.

Dupuis had stated that he had printed the book for the use of the Tamil seminarians. There is a plea which states in the book 'dear students please understand the use and good of the learning Latin and read this book. Kindly pray to the Lord for those who prepared and brought it out'.

There is further a mention that limited copies of the book would only be found marketable in Tamil country. Hence the price of the book had been fixed at two rupees. However, in the larger interest of the public for use, the price of this book had been greatly reduced for local sales and the price had been fixed at one rupee. This would enable the people of Pondicherry to buy, use and benefit. This is the first book written in Tamil language on Latin grammar and also got printed for the first time.

Ilakkana Surukkam: The Elements of Tamil Grammar in Tamil Printed in 1845

The MEP missionaries very much desired to use a printed Tamil grammar book in the seminary at Pondicherry instead of using the customary palm-leaf Tamil manuscript. This they planned so that the seminarians may obtain the required skill over their own mother-tongue. The missionaries had contacted the noted Tamil scholar Savarimuthu Pillai the resident of Santhome of Mylapore for composing this book. Accordingly he composed the elements of Tamil grammar in prose since Tamil grammar was only available in poetry. This printed book was titled Ilakkana Surukkam and the same was published from the Catholic press in Pondicherry with the permission of the MEP superiors in 1845. According to the title page, we find the name of the author Savari Muthu Pillai who composed at the request and sanction of the MEP missionaries. The printed book contains 182 pages in total.[10] The erratum had been found at pp. 181-2.

The preface states that the book is composed for all those who wish to learn Tamil grammar very easily in Tamil prose including the children since Tamil grammatical works had been available in poetry and also in commentaries. As this book is very small, the porul ilakkanam and the ani ilakkanam had been omitted. Thus the book deals with only yezhuthu, (letters/scripts) sol (word), and yappu (prosody) the three parts of the grammar. It has been stated that the book has been written after conducting serious research and also after going through the various Tamil grammar works existed.

The first part Yezhuthathigaaram deals with yezhuthiyal, punariyal, uyir ittru punarchi, mey ettru punarchi, pothu punarchi, seyyul vigaarangal and vadamozhi sandhi in the book.[11] The next part Solathigaaram deals with peyar iyal, peyar sol, vettrumai, vinai iyal and some exampales of viguthi, besides pothuviyal, idaisol iyal and urisol iyal.[12] The third part Yappathigaaram deals with asai, seer, thazhai, adi, thodai. Then we find Paavum paavinamum, Venpaa ilakkanam, Asiriyappa ilakkanam, Kalipaa ilakkanam, Vanjipaa ilakkanam, Marutpaa ilakkanam, Paavinam, Thurai ilakkanam, Thazzh isai ilakkanam and Viruththa ilakkanam.[13]

Finally we find the list of 75 prabhandams and how they have been composed and written with what rules. It also mentions in detail how their grammar should be in Tamil poetry.[14] The prabhandams listed are as follows:

1. Pillai kavi
2. Kalambagam
3. Bharani
4. Agapporul kovai
5. Varukkak kovai
6. Mummanik kovai
7. Atta mangkalam
8. Anga maalai
9. Rettai mani maalai
10. Inay mani maalai
11. Nava mani maalai
12. Naanmani maalai
13. Naama maalai
14. Palasandha maalai
15. Panmani maalai
16. Mani maalai
17. Pugalchi maalai
18. Peru magilchi maalai
19. Varukka maalai
20. Meykrithi maalai
21. Kaapu maalai
22. Venin maalai
23. Vasantha maalai
24. Karpu maalai
25. Urpava maalai
26. Thaanai maalai
27. Mummani maalai
28. Thandaga maalai
29. Veera vetchi maalai
30. Vettri karandhai maalai
31. Por kezhu vanji
32. Varalaattru vanji
33. Serukkazhla vanji
34. Kaanji maalai
35. Notchi maalai
36. Ulzhinhai maalai
37. Thumbai maalai
38. Vaagai maalai
39. Vathorana manjari
40. Yen Cheyyul
41. Thogai nilai cheyyul
42. Oli andhaathi
43. Pathitrr andhaathi
44. Nootr andhaathi

45. Ulaa
46. Orupaa oru paththu
47. Irupaa irupathu
48. Aatrupadai
49. Kanpadai nilai
50. Thuyil idai nilai
51. Peyar innisai
52. Ur innisai
53. Peyar nerisai
54. Ur nerisai
55. Ur venpaa
56. Vizhlakku nilai
57. Pura nilai
58. Kadai nilai
59. Thasaanga paththu
60. Thasaanga thayal
61. Arasan viruththam
62. Maathathi kesam
63. Kesaathi paatham
64. Nayanap paththu
65. Alangaara panchagam
66. Mangkala vazhllay
67. Naarpathu
68. Kuzha magan
69. Thaandagam
70. Pathigam
71. Sathagam
72. Seviyarivu rrovu
73. Vayurai vaazhththu
74. Pura nilai vaazhththu
75. Voosal

The book prepared and printed was aimed to enhance the Tamil language skill by the seminarians in Pondicherry and also to serve the need of common man. The book had also appeared as second edition from the same Catholic press in Pondicherry in 1860 owing to its demand.[15]

Ilakkana Noolaathaaram: The Basics of Tamil Grammar in Tamil for the Use of Children Printed in 1849

The MEP missionaries in Pondicherry felt the need to print a Tamil grammar book in Tamil for school-going children. In the past, kids were trained to use the palm-leaf manuscript in Tamil in the traditional way. The missionaries felt the need to use the printed text book. Hence Mousset the professor in the seminary in Pondicherry decided to compose a book by using the existing works in Tamil besides using the Thonnul Vizhlakkam composed by Beschi. Mousset completed work and the book titled Ilakkana Noolaatharam was printed at the Catholic press in Pondicherry by the order of the MEP Superiors in 1849.[16] The title page mentions that the book was composed by one of the missionaries. According to missionary letter correspondences it is learnt that Mousset was the author of this book. The total pages of the book are 80. It is found mention in the book that the details of the several rules of Tamil grammar could please be found in Yappathigaaram and Thonnul (vide p. 80). Hence it is certain Beschi's Thonnul Vizhlakkam had been largely consulted and used in the preparation of Ilakkana Noolaatharam.

There is a short preface in Tamil and it states that grammar is very important for writing and speaking the language properly without error. It explains how letters formed words and words formed the verses. The author of the book says that the book is so titled as Ilakkana Noolaathaaram because it contains the primary rules of Tamil grammar in brief. Those who read this book and learn Tamil grammar well would later easily be able to understand and discover the contents of other books like Thonnul (composed by Beschi) and Ilakkana Surukkam (printed at Pondicherry in 1845).

The book contains only yezhthuathigaaram and solathikaaram. The first part of the book deals with yezhuthathikaaram and explains punar iyal.[17] Then we find the second part dealing with solathikkaaram and explains peyar iyal, vinai iyal, idaissol iyal, urisoliyal, and thodarpiyal.[18]

The purpose of writing this book is neatly spelled out by the author and he says that his target audience had been the Tamil school children and in order to make them understand Tamil grammar this book is written in prose. He says that in the reading of this book one can easily find that those parts of the Tamil grammar which had not been usually provided. Even those that had been omitted in Kodunthamizh (of Beschi) have been explicitly printed and given in very smaller founts in this book. Hence it appears as a comparative grammar of Tamil printed books.

Grammaire Française-Tamoule: The Tamil Grammar in French Authored by Dupuis and Printed in 1863

Dupuis in Pondicherry printed the Tamil grammar in Latin of Beschi for learning Tamil by the Europeans. But there was no grammar book for learning Tamil by the French audience. In order to remedy this Dupuis planned to compose such a work and he successfully executed. The book titled Grammaire Française-Tamoule appeared in print at the press of the missionaries in Pondicherry in 1863. The title of the book mentions French-Tamil grammar or rules of vulgar Tamil and high language called poetic Tamil exposed by a pro-vicar apostolic of the congregation of foreign missionaries, i.e., Louis Savinien Dupuis.[19] The printed book had a total of 554 pages.

Dupuis had used the earlier works of the French missionaries in Pondicherry who wrote French-Tamil Grammars during the eighteenth century. Pierre de la Lane born in 1669 who came to India in 1703 a Jesuit, learnt Tamil in 1703 while stationed at Takkolam situated in the hinterland of Pondicherry he composed Grammaire Française-Tamoule (French-Tamil grammar) in 1728.[20] After his death in 1746 his grammar manuscript was sent from Pondicherry to France and it is now preserved at the Bibliothèque Nationale, Paris.[21] Pierre de la Lane had given details in French on Tamil alphabets, their pronunciation, and the grammar of Tamil at length by explaining the kinds of nouns, adjectives, pronouns, verbs, conjugations, tenses, imperatives, infinitives, adverbs, preposition, conjunction and interjection.

Further, Dupuis had also used the French-Tamil grammar of Dominique de Valence (1696-1778), the superior of the Capuchins in Pondicherry who learnt Tamil had composed grammaire française-tamoul (French-Tamil grammar) in 1734.[22] Valence had given details in French on *preliminaire sur l'alphabet et la pronunciation, grammaire de la language tamoule* explaining the kinds of nouns, adjectives, pronouns, verbs, conjugations, tenses, imperatives, infinitives, adverbs, preposition, conjunction and interjection. It must be admitted that Dupuis had used the above said two works and enlarged his work while composing a new grammaire française-tamoule. It may be stated that the grammar manuscripts of the two French missionaries had remained in manuscript form without printing. However, the credit goes to Dupuis for printing the grammaire française-tamoule for the first time in 1863 since he had expanded and revised the work of the earlier missionary writers.

Dupuis in his Tamil grammar in French printed had given a preface. He mentions that ever since that he brought out the publication of Beschi's Tamil grammar in Latin in 1843 it gave an impetus for executing this new work. The revision of Beschi's grammar had been carried out by him and so the present book had been made in a different style with rules of vulgar Tamil and the high language called poetic Tamil with additions though Beschi was the grand authority of Tamil grammar. Dupuis said that he considered it necessary to have a Tamil grammar in French for use of the people in the French colony of Pondicherry so that various functionaries of the government, business and commercial establishments could benefit. Dupuis stated that there is a great demand for such a grammar of this kind since the emigration of the coolies to other French colonies like Bourbon and Martinique had taken place from Pondicherry where the Tamils worked and those French stationed there could also learn Tamil through this book.[23] Further, Dupuis added that The Tamil course offered in the colonial college in Pondicherry was another necessity for printing this grammar book since this text book could be used for teaching and learning. The MEP society and the evangelical workers in India used Tamil grammar in Latin and the great majority of the apostolic missionaries working in the various vicariates including Sri Lanka were French speaking individuals and so in order to facilitate their studies in Tamil, a grammar of this kind was found absolutely necessary. In fine, Dupuis said that there is a total necessity for serving them and helping in the studies of the Tamils in the overseas French colonies as well as to learn French too. Therefore under these circumstances, Dupuis was firm in his opinion to compose and publish this grammar.[24]

Dupuis at the outset mentioned the plan of the grammar book.[25] It is divided into three parts. Part one deals with letters, pronunciation, diverse kinds of words, and the manner of declensions and conjugations. Part two of the book deals with the syntax rules for union of words, formation of phrases, rules for orthography, formation and employment of diverse appellative names, the manner of employing different times and moods of verbs useful for learning the language. Part three of the book exposed the rules for versification of Tamil and diverse kinds and species of poetry. Dupuis also provided nine important points regarding Tamil grammar and the manner of understanding on the use of article in Tamil, masculine and feminine gender, participles, conjugation, words, adjectives and adverbs, conjugations and phraseology of Tamil.[26]

While composing this new work Dupuis said that this book was especially directed and meant for the use of the Europeans. But he explained the possible methods of grammars used in Europe were found useful along with Beschi's work of erudition and also of various others to overcome the lacunae found in the earlier writings. Therefore he applied his mind and corrected the mistakes and defects. Some syntax had been explicated or other developments could be found in this book.

To conclude it may be said that Dupuis took pains to read earlier grammar works. He understood the significance and the need for revision and composition of new books. He also took the help of others like Savarimuthu Pillai and Louis-Marie-Mousset.

Notes

1. Archives de la Société des Missions Etrangères de Paris (hereafter AMEP), Correspondance Pondichéry-Paris, Lettres, vol. 1000, see the letter of L.S. Dupuis from Pondicherry dated 16 March 1844 to Jean Tesson in Paris.
2. AMEP, Correspondance Pondichéry-Paris, Lettres, vol. 1000, see the letter of Mousset dated 19 April 1845 to Jean Tesson.
3. Bibliothèque Asiatique des Missions Etrangères de Paris (hereafter BAMEP), Piranchu Ilakkana Noolsurukkam, TAM1, pp. 168-170.
4. Ibid., pp. 154-9.
5. Ibid., pp. 159-67.

6. AMEP, Correspondance Pondichéry-Paris, Lettres, vol. 1000, see the letter of L.S. Dupuis from Pondicherry dated 16 March 1844 to Jean Tesson in Paris.
7. AMEP, Correspondance Pondichéry-Paris, Lettres, vol. 1000, see the letter of Mousset dated 19 April 1845 to Jean Tesson.
8. Madurai Province Jesuit Archives (hereafter MPJA), Shenbaganur, 221/133, Lathin Ilakkana Noorsurukkam.
9. Ibid., pp. 225-7.
10. MPJA, 221/132, Ilakkana Surukkam.
11. Ibid., pp. 8-58
12. Ibid., pp. 59-133.
13. Ibid., pp. 134-64
14. Ibid., pp. 165-79.
15. BAMEP, Ilakkana Surukkam, second edition, Pondicherry, 1860.
16. BAMEP, TAM2, Ilakkana Noolaatharam; See also, AMEP, 03891, Ilakkana Noolaatharam.
17. BAMEP, TAM2, pp. 1-26.
18. Ibid., pp. 27-80.
19. BAMEP, C3166, *Grammaire Française-Tamoule ou les regles du tamoul vulgaire, celles de la langue relevee, et celles de la poesie tamoule se tuouvent exposees au long par un pro-vicaire apostolique de la congregation des missions–étrangères*, Pondichéry, Imprimerie des missionaries apostoliques de la dite congregation 1863 avec l'approbation des superieurs.
20. *Lettres édifiantes et curieuses, écrites des missions étrangères par quelques missionaires de la Compagnie de Jésus* (hereafter Lettres édifiantes et curieuses), 34 vols., Paris, 1702-1777, vol. X, p. 397; Lettres édifiantes et curieuses, vol. X, p. 42. The grammar attributed to Pierre de la Lane is titled 'Grammaire pour apprendre la langue Tamoul diverse vulgairement le Malabar, 1728'.
21. Bibliothèque Nationale, Paris (hereafter BNP), MSS Indien, no.189, fls.1-92;
22. Det Kongelige Bibliotek (hereafter KB), Kobenhavn, Oriental Department, MSS. Cod Tamul 30 (UB). The grammar was composed in 1734 and the year is neatly mentioned. BL, London, European Manuscripts (hereafter MSS EUR), E.121, *Dictionnaire et grammaire francais tamoul*. The dictionary is first mentioned which was completed in the year 1743. Hence there is no confusion between the grammar composed in the year 1734 and the dictionary that was completed in 1743.
23. Société Archeologique de Sens (hereafter SAS), *Grammaire Française-Tamoule, ou les relges du Tamoul vulgaire, celles de la langue relevee et par celles de la poesie Tamoule se trouvent exposees au long par un provicaire Apostolique de la Congregation des Missions Etrangères*, Pondichéry, Imprimerie des Missionnaires Apostoliques de la dite Congregation, 1863, see the preface, p. 5.
24. Ibid., p. 6.
25. Ibid., p. 7.
26. Ibid., pp. 9-13.

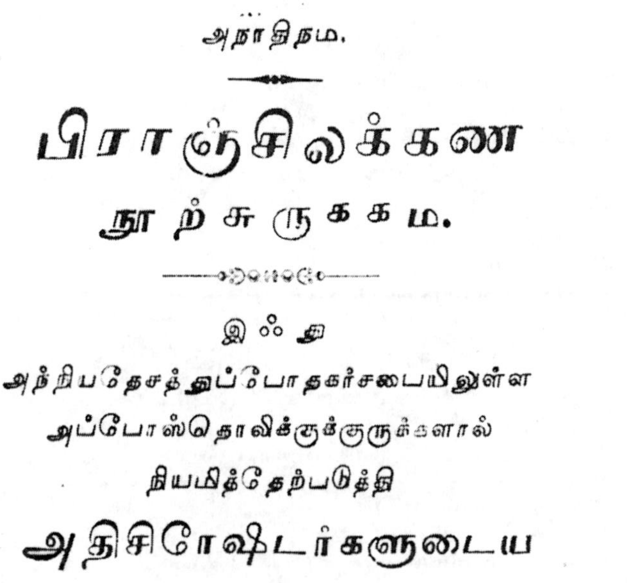
அநாதிநம.

பிராஞ்சிலக்கண

நூற்சுருக்கம.

இஃது

அந்நியதேசத்துப்போதகர்சபையிலுள்ள

அப்போஸ்தொலிக்குக்குருக்களால்

நியமித்தேற்படுத்தி

அதிசிரேஷ்டர்களுடைய

உத்தாரத்தின்படியே

௲௮௱௪௰௫

புதுவையிற்சென்மவிராக்கினிமாதாகோயிலைச்சேர்ந்த

அச்சுக்கூடத்திற்பதிப்பிக்கப்பட்டது.

5.1. The title page of *Piranch Ilakkana Noorsurukkam* printed in 1845 (Courtesy: Bibliothèque Asiatique des missions étrangères de Paris).

வினைச்சொல்லியல். ௮௩

Nous nous baignerons குளிப்போம்.
Vous vous baignerez குளிப்பீர்கள்.
Ils se baigneront குளிப்பார்கள்.

இறப்பிடையெதிர்காலம்.

Je me serai baigné குளித்திருப்பேன்.
Tu te seras baigné குளித்திருப்பாய்.
Il se sera baigné குளித்திருப்பான்.
Nous nous serons baignés குளித்திருப்போம்.
Vous vous serez baignés குளித்திருப்பீர்கள்.
Ils se seront baignés குளித்திருப்பார்கள்.

உடன்பாட்டுவிதம். நிகழ்காலம்.

Je me baignerais குளிக்கத்தக்கவனாயிருப்பேன்,
Tu te baignerais குளிக்கத்தக்கவனாயிருப்பாய்,
Il se baignerait குளிக்கத்தக்கவனாயிருப்பான்,
Nous nous baignerions குளிக்கத்தக்கவர்களாயிருப்போம்,
Vous vous baigneriez குளிக்கத்தக்கவர்களாயிருப்பீர்கள்,
Ils se baigneraient குளிக்கத்தக்கவர்களாயிருப்பார்கள்.

இறந்தகாலம்.

Je me serais baigné குளித்திருப்பேன்,
Tu te serait baigné குளித்திருப்பாய்.
Il se serait baigné குளித்திருப்பான்.
Nous nous serions baignés குளித்திருப்போம்.
Vous vous seriez baignés குளித்திருப்பீர்கள்,
Ils se seraient baignés குளித்திருப்பார்கள்.

வேறுபிரகாரமாவது.

Je me fusse baigné. Nous nous fussions baignés.
Tu te fusses baigné, Vous vous fussiez baignés.
Il se fût baigné. Ils se fussent baignés.

ஏவல்விதம்.

(தன்மையிடமில்லை.)
Baigne-toi குளி.
Qu'il se baigne குளிப்பானாக.
Baignons-nous குளிப்போமாக.
Baignez-vous குளியுங்கள்.
Qu'ils se baignent குளிப்பார்களாக.

5.2. Page 83 of Piranch Ilakkana Noorsurukkam explaining conjuga-tions (Courtesy: Bibliothèque Asiatique des missions étrangères de Paris).

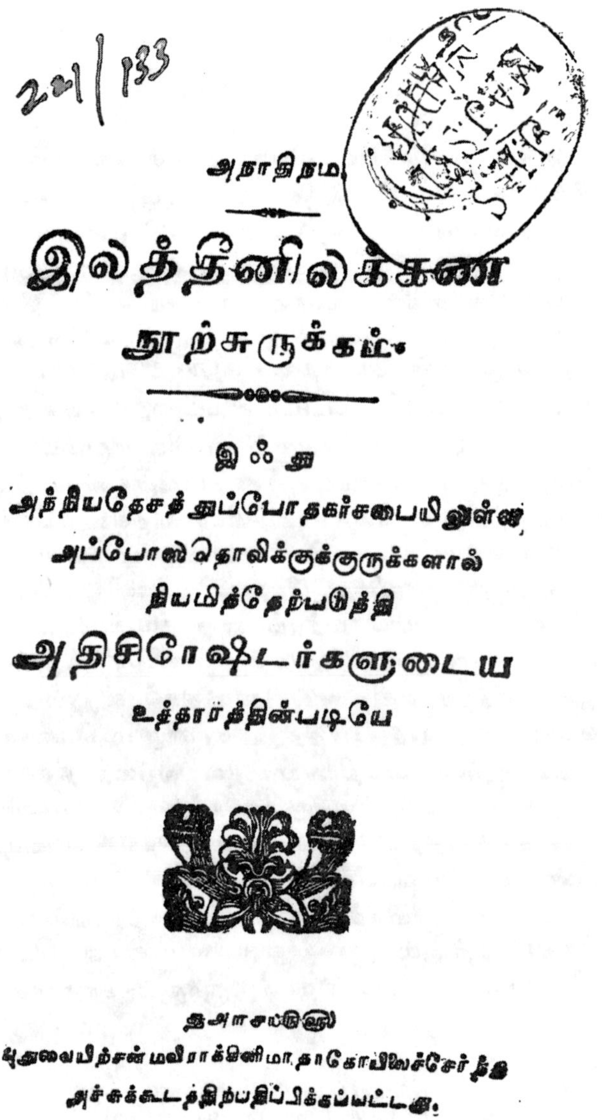

அநாதிநம

இலத்தினிலக்கண

நூற்சுருக்கம்.

இஃது

அந்நியதேசத்துப்போதகர்சபையிலுள்ள

அப்போஸ்தொலிக்குக்குருக்களால்

நியமித்தேற்படுத்தி

அதிசிரேஷ்டர்களுடைய

உத்தாரத்தின்படியே

௲௮௱௪௰௫ ஸல்

புதுவையிற்சன்மவீராக்கினிமாதாகோவிலைச்சேர்ந்த

அச்சுக்கூடத்திற்பதிப்பிக்கப்பட்டது.

5.3. The title page of Lathinilakkana Noorsurukkam printed in 1845 (Courtesy: Madurai Province Jesuit Archives)

பாயிரம.

இலக்கணவிலக்கியங்களையு மற்றச்சகலகல்விசா
ஸ்திரங்களையும விளக்குகிறவைகளும், எத்தேசங்
களிலு மெக்காலங்களிலுங் கற்றோரெல்லார்க்கும்
பிரயோசன மாதிரிகையானவைகளுமாகிய வெ
ண்ணிக்கைக்குள்ளடங்காத உத்தம நூற்க ளெ
ழுதப்பட்டபாஷையா யிருப்பதினாலும, வேதசா
ஸ்திரத்துக்குஞ் சாதாரணசத்திய வேத தேவா
ராதனைக்கு நியமிக்கப்பட்டமையினாலும், இலத்தீ
ன்பாஷை பூலோகமெங்கும் நிகரில்லாமுக்கியம்
பெற்றது. மீளவுங் கடவுள்தார்மதிருவுளம்பற்றி
னவேதத்திலும் வேதசாஸ்திரிகள் தேவவாக்கியங்
களை விளக்கினவியாக்கியானங்களிலும் வேதபார
கர்தந்தருளிய அமிர்தப்பிரசங்கநூற்களிலு மிவை
முதலிய வுத்தமபிரபந்தங்களிலு மடங்கியபரம
வுணர்ச்சிகள் ஊறுகிற சுத்தவூறணியா யிருப்ப
துமன்றிச் சகலவித கல்விசாஸ்திரங்களிலும் புல
மைமுதலிய வித்தியாதத்துவங்களிலும் மேலாய
திறமையும் சத்தியநியாயமுதலியவற்றை நிச்சயி
த்து இருதயத்தைவிளக்குகிறவாய்ச்சாலகமுமுண்
டாவதற் கெத்தனமாயிருத்தலு மிலத்தீன்பாஷை
யின் மாட்சிமையாமே.

ஆதலால் பூர்வீகந்துவக்கி எத்தேசத்திலும் எச்
சாஸ்திரத்திலும் ஞானிகள் கல்விமான்கள் பெரி
யோர்கள் முதலியசகலகற்றோருக்கும்-சாதாரண
பாஷையானதிதுவே. ஆகையால் இத்தேசத்திலு

5.4. The preface of Lathinilakkana Noorsurukkam (Courtesy: Madurai Province Jesuit Archives).

அ* இலத்தீன்பாஷையின்

Sma sme smi smo smu | Squa sque squi squo squu
ஸம ஸமெ ஸமி ஸமொ ஸமு | ஸகு அஸகு எஸகு இஸ்கொ ஸ்க்கு

இலத்தீனபாஷையைவாசிக்க அதிகவெளிதாய் அப்பியாசமுண்டாகும்பொருட்டுச் சிலஇலத்தீன்மந்திரங்களை இலத்தீன் எழுத்திலுந் தமிழ் எழுத்திலு மிங்கேதருகிறோம். F z எனகிற வெழுத்துக்களுக்குப்பதிலெழுத்துத்தமிழிலில்லாமையால் இதிலேயவைகளைத் தமிழிலுமபிரயோகிப்போம்.

In nomine Patris et Filii et Spiritus
இன் னொமினெ ப்பாத்திறிஸ் எத்த் பிலியி எத்த் ஸ்பிரித்துஸ்
sancti. Amen.
சங்க்த்தி ஆமென்

Oratio Dominica.
ஒறாசியொ தொமினிக்கா

Pater noster qui es in cœlis, sanctifice-
பாத்தேர் நொஸ்தேர் குஇ எஸ் இன் செலிஸ் சங்க்த்திபிசெ
tur nomen tuum: adveniat regnum tuum:
த்துர் நோமென் த்துஉம் அத்வெனிஅத்த் றெக்நொம் த்துஉம்
fiat voluntas tua sicut in cœlo et
பிஅத்த் வொலொந்த்தாஸ் த்துஅ சிக்குத்த் இன் செலொ எத்த்
in terra. Panem nostrum quotidianum da
இன் தெறு ப்பானெம் நொஸ்திரும் கொத்திதிஆனும் தா
nobis hedie: et dimitte nobis dibita nostra,
நோபிஸ் ஒதிஎ எத்த் திமித்தெ நோபிஸ் தெபித்தா நொஸ்திரு
sicut et nos dimittimus debitoribus nostris:
சிக்குத்த் எத்த் நோஸ் திமித்திமுஸ் தெபித்தோரிபுஸ் நொஸ்திரிஸ்
et ne nos inducas in tentationem: sed
எத்த் நெ நோஸ் இந்துக்காஸ் இன் த்தென்த்தாசியோனெம் செத்
libera nos a malo. Amen.
லிபெரு நோஸ் அ மாலொ ஆமென்

Salutatio Angelica
சலுத்தாசியொ அந்ஜெலிக்கா.

Ave, Maria, gratia plena, Dominus tecum:
ஆவெ மரியா கிருசியா ப்பிளெனு தொமினுஸ தெகொம்
benedicta tu in mulieribus, et benedictus
பெனெதிக்த்தா த்து இன் முலியெரிபுஸ எத்த் பெனெதிக்த்துஸ

5.5. Prayers in Latin and Tamil transcription printed in Lathinilakkana Noorsurukkam (Courtesy: Madurai Province Jesuit Archives).

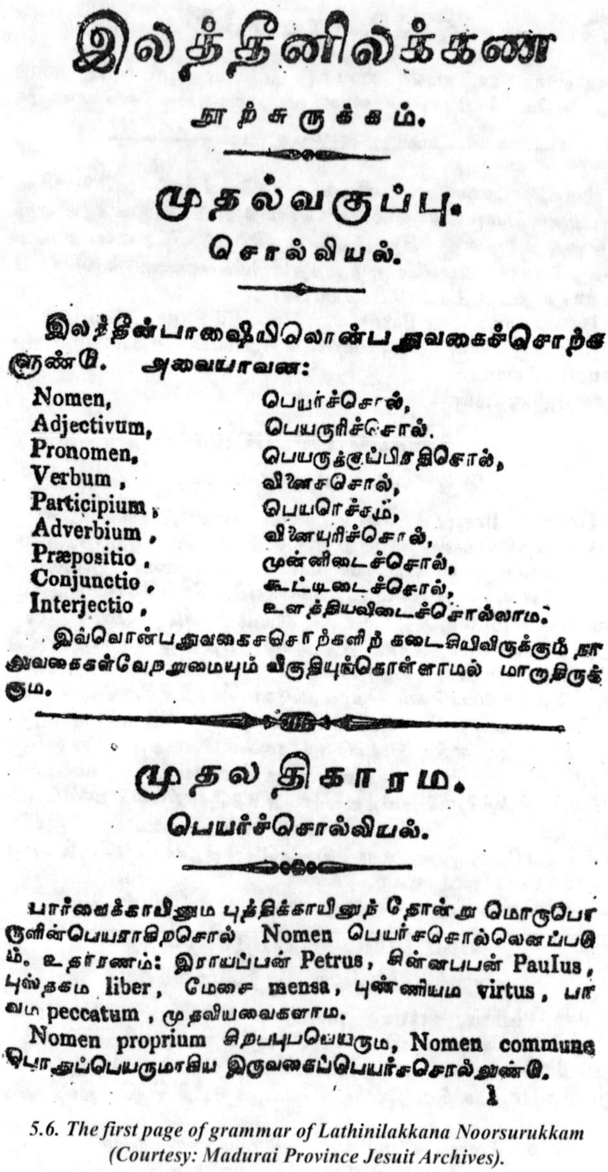

இலத்தீனிலக்கண

நூற்சுருக்கம்.

முதல்வகுப்பு.

சொல்லியல்.

இலத்தீன்பாஷையிலொன்பதுவகைச்சொற்களுண்டு. அவையாவன:

Nomen, பெயர்ச்சொல்,
Adjectivum, பெயருரிச்சொல்.
Pronomen, பெயருக்குப்பிரதிசொல்,
Verbum, வினைச்சொல்,
Participium, பெயரெச்சம்,
Adverbium, வினையுரிச்சொல்,
Præpositio, முன்னிடைச்சொல்,
Conjunctio, கூட்டிடைச்சொல்,
Interjectio, உளத்தியலிடைச்சொல்லாம்.

இவ்வொன்பதுவகைச்சொற்களிற் கடைசியிலிருக்கும் நாலுவகைகள்வேற்றுமையும் விகுதியுங்கொள்ளாமல் மாறாதிருக்கும்.

முதலதிகாரம்.

பெயர்ச்சொல்லியல்.

பார்வைக்காயினும் புத்திக்காயினுந் தோன்றுமொருபொருளின்பெயராகிறசொல் Nomen பெயர்ச்சொல்லெனப்படும். உதாரணம்: இராயப்பன் Petrus, சின்னப்பன் Paulus, புஸ்தகம் liber, மேசை mensa, புண்ணியம் virtus, பாவம் peccatum, முதலியவைகளாம்.

Nomen proprium சிறப்புப்பெயரும், Nomen commune பொதுப்பெயருமாகிய இருவகைப்பெயர்ச்சொலுண்டு.

1

5.6. The first page of grammar of Lathinilakkana Noorsurukkam (Courtesy: Madurai Province Jesuit Archives).

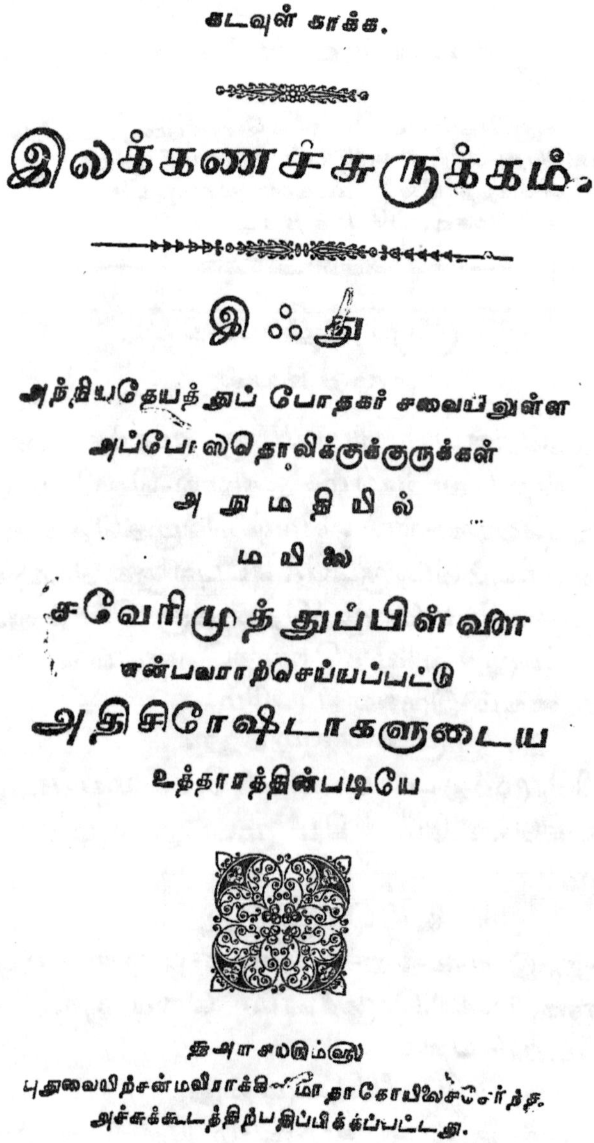

5.7. The title page of Ilakkana Surukkam printed in 1845 (Courtesy: Bibliothèque Asiatique des missions étrangères de Paris).

இலக்கணச்சுருக்கம்.

கடவுள்வணக்கம்.

வெண்பா.

உலகுயிர்கள் பல்பொருள்களொண்டமறை மற்றெல்லா
நிலைபெறச்செய்தாளுநிருபன் - கலைமதித்தாட்
கன்னியெந்தாயீன்ற கடவுண்மலரடியைச்
சென்னிமிசைவைப்போந் தினம்.

எழுத்ததிகாரம்.

எழுத்தியல்.

எழுத்தென்பது மொழிக்கு முதற்காரணமாயிருக்கின்ற வொலியாம். ஆதலால்-செவிப்புலனாகிய. இவ்வோசையை-ஒலிவடி வெழுத்தென்றும்-இதனையறியும்பொருட்டுக் கட்புலனாக வெழுதப்படுவதை வரிவடிவெழுத்தென்றுஞ் சொல்லுவர்.

அவ்வெழுத்து முதலெழுத்தெனவும்-சார்பெழுத்தெனவும்-இருவகைப்படும்.

முதலெழுத்து.

உயிரெழுத்துப் பன்னிரண்டும்-மெய்யெழுத்துப் பதினெட்டும்-ஆகிய முப்பதும்-முதலெழுத்தாம்.

உயிரெழுத்து.

அ-ஆ-இ-ஈ-உ-ஊ-எ-ஏ-ஐ-ஒ-ஓ-ஔ - இப்பன்னிரண்டும்-உயிரெழுத்தாம். இவை-ஆவியெனவும் பெயர்பெறும்.

மெய்யெழுத்து.

க்-ங்-ச்-ஞ்-ட்-ண்-த்-ந்-ப்-ம்-ய்-ர்-ல்-வ்-ழ்-ள்-ற்-ன் - இப்பதினெட்டும்-மெய்யெழுத்தாம். இ

5.8. A page showing grammar in Ilakkana Surukkam
(Courtesy: Bibliothèque Asiatique des missions étrangères de Paris).

இலக்கணநூலாதாரம்.

கடவுள்-துணை.

முகவுரை.

இலக்கணமென்பது [illegible]

எழுத்ததிகாரம்.

முதற்பிரிவு. — எழுத்தியல்.

தமிழெழுத்து முதலெழுத்தெனவுஞ் சார்பெழுத்தெனவு மிருவகைப்படும்.

முதலெழுத்து. — உயிரெழுத்துப் பன்னிரண்டும் மெய்யெழுத்துப்பதினெட்டுமாகிய இம்முப்பதும் முதலெழுத்தெனப்படும்.

D க

5.9. The page showing the preface of Ilakkana Noolaathaaram (Courtesy: Bibliothèque Asiatique des missions étrangères de Paris).

DICTIONAIRE ET
GRAMAIRE
FRANCOIS TAMOVL

Compoſé par le Reuerend Pere
Dominique, Miſſionaire Ap-
oſtolique et Superi-
eur a Pondichery.

PONDICHERY

MDCCXLIII

5.10. The title page of the manuscript of Grammaire Française-Tamoule of Dominique de Valence of 1743 (Courtesy: British Library, London).

GRAMMAIRE
FRANÇAISE-TAMOULE,
ou
LES RÈGLES DU TAMOUL VULGAIRE,
CELLES DE LA LANGUE RELEVÉE,
ET CELLES DE LA POÉSIE TAMOULE
SE TROUVENT EXPOSÉES
AU LONG
PAR UN PROVICAIRE APOSTOLIQUE
DE LA CONGRÉGATION
DES MISSIONS - ÉTRANGÈRES.

PONDICHÉRY,
IMPRIMERIE DES MISSIONNAIRES APOSTOLIQUES
DE LA DITE CONGRÉGATION.
1863.
Avec l'approbation des Supérieurs.

5.11. The title page of Grammaire Française-Tamoule printed in 1863 (Courtesy: Bibliothèque Asiatique des missions étrangères de Paris).

Part II

Missionaries, Linguistics and Tamil Multiple Dictionaries in Discourse and Difference

6

The Birth of Bi-Lingual and Tri-Lingual Dictionary in Tamil: Missionary Enterprises and the Manuscripts Prepared and Circulated 1549-1773

The beginning of Tamil dictionary could be gleaned directly from the classical work of Tholkaapiyam, where alphabetical arrangements of words appear not strictly or logically. There is a chapter on qualifying words or qualifiers which deals with lexical matter.[1] There are two more chapters, while one deals with particles and the other on literary usage with names of masculine gender, feminine gender which contains matters of lexicographical interest.[2] In another chapter on miscellaneous we find details of the classification of words into four groups, such as common native words, literary words, words of the then foreign origin (*thisai-sol*) and words of Sanskrit origin (*vada-sol*) in Tamil language.[3] Thus we find the rough classification of the sources of Tamil vocabulary in the ancient period.[4] We cannot certainly say that there is dictionary in Tholkaapiyam, but what we find is that the seed for the growth of dictionary being present.

Following the age of Tholkaapiyar, Tamil country witnessed intense religious propaganda through Buddhism, Jainism, Saivism and Vaishnavism. Sanskrit language at that time had a powerful influence and increased with its words abundantly the vocabulary of Tamil. According to the then desired needs, some works of reference, such as Dhivaagaram, a dictionary came to be composed in the middle of the eighth century. The author of this work is Dhivaagarar. The work contains ten sections and deals with class–vocabularies, divided into sections according to subject matter where we find the names of gods and deities, heavenly bodies, names of ranks, orders of men, parts of the body, names of birds, beasts, insects, plants, trees, places, countries, tools and weapons. In the eleventh part we find one word with many meanings and the twelfth part deals with five senses and seven births. It approximately contains 9500 words and there is a reference in this work to the Chalukya rulers.[5]

The next class-vocabulary in Tamil called Pingalanthai appeared between eighth and thirteenth centuries and the author was Pingalar, son of a certain Dhivaagarar. It has a vocabulary enlarged in many directions and first found mentioned in Nannul (vide Sutra 460), a Tamil grammar of the thirteenth century. Here we have ten parts and the tenth part deals with the eleventh part of Dhivaargarar giving details of one word with multiple meanings. The rest nine parts are expanded version of the rest of the things found in the work of Dhivaagaram. There are approximately 14,700 words in Pingalam.

Besides these above works, there has been also many wordbooks such as the work of Ilampuranar of the eleventh century, the great commentator on Tholkaapiyam where we find a Nikandu Sutra (vide sol-athikaaram, 390) of which the source is not yet traced. Similarly Yapperunkala Viruththi-urai also gives the various sources of the word apparently from a sutra, not found in any of the existing nikandus.[6] Thus works such as Tholkaapiyam, Dhivaagaram, and Pingalam paved the way for the origin and development of Tamil lexicons. They had Tamil words in alphabetical order and their meanings were also given. Tamil lexicography had thus out-grown the crude form of class-vocabularies.

Chudamani nikandu, another work was composed in 1520 AD and its author was Mandala Purutan, a Jain. He in his prefatory stanzas made it clear that he was a chieftain of Veerai and the disciple of Gunabhadra, a Jain. A reference is made to Krishnadevaraya (AD 1509-1529) the ruler of Vijayanagara in this work. This Chudamani nikandu became famous for it was helpful to memorize easily. Bartholomäus Ziegenbalg the Protestant missionary in Tranquebar in 1709 mentions this Chudamani nikandu as a poetical book that had been like the Dhivaagaram and it is rich in words but there is one essential difference, while the other work was written in prose, this nikandu was written entirely in verse. The author he mentioned was called Veera Mandalavar, who lived at the same time as Dhivaagarar and composed this work on the basis of the Dhivaagaram.[7]

Kaarigai another Tamil text is also mentioned by Ziegenbalg and according to him it shows sixteen different types of verses, explaining how, according to the rules of grammar, one can modify verses so that they suit different melodies just as it happens in the case of European songs. The author of this book was Amirtha Sagaran who lived some hundred years ago according to Ziegenbalg (i.e., around 1600 AD). He was a hermit and he learnt this art from a great prophet who, together with some thousand other prophets, was still living inside a mountain. This great man by the name of Agathiyar once drank the whole sea empty and discharged it again through his urine; this, the Tamils think, is the reason why sea-water tastes salty. The mountain where this prophet lived was called Pothigai Malai; it is fifteen days' journey from Tranquebar. Kaarigai was first written in verse by the author himself, then another poet wrote the commentary; therefore under every verse one can find its appropriate explanation. Nevertheless it is one of the most difficult books the Tamils have and can only be found in the possession of poets.[8]

Agarathi nikandu was another important lexicographical work composed by Revana Siddhar of Chidambaram (then known as Puliyur). In his paayiram (preface) we find mention of saka year 1516 (1594 AD) in which he had composed. The author had arranged Tamil words (A. AA, E, EE) in alphabetical order.[9] This was the first work to bear the name agaraathi (dictionary) in Tamil. Reference and the alphabetical arrangement of Tamil words had evolved, though it had not advanced beyond the second letter of the Tamil word. Such was the stage of mono-lingual vocabulary in Tamil just at the time, after the advent of the Portuguese in Tamil country. Thus nikandus paved way for the making of dictionaries. They served as the basis for tools for dictionaries and were composed in Tamil poetry.

As the missionaries from Europe came and had a purpose to study Tamil language they found the Tamil language almost presented a different spectacle unlike the languages of Europe. Every single element of difference in learning Tamil had attracted their attention. They themselves wanted to get well versed in Tamil, so that they could in turn reach out to the masses with their message of Christianity. They were able successfully to learn Tamil grammar and also introduced some innovations in the field to reach the audience in the West. At this time they desired to enhance the vocabulary of words to master the language to communicate fluently with Tamils. Therefore they prepared bi-lingual dictionaries in various European languages.

The Lexicographical Attempt by the Portuguese Traders in Malabar and the Early Beginnings in Linguistic Terms, 1498-9

The early evidence in Portuguese on the production of a dictionary in Malayalam language on the Malabar Coast is traced. The report of the first voyage of Vasco da Gama by Alvaro Velho contains a linguistic appendix found at the end of the chapter under the epitaph '*estba e' a linguagem de*

Calicut' and it includes 123 terms generally of nouns, adjectives and verb forms besides fifteen proper names in Portuguese and Malayalam. This is the first rudimentary vocabulary (Portuguese-Malayalam vocabulary) but lacks a pre-established alphabetical or thematic order. Thus Malayalam was the first language in which Portuguese gathered words from oral production and compiled a vocabulary within seven weeks of period when they were in Calicut. The vocabulary words included dealing with parts of the body, dress, human actions, weapons, geographical/astronomical terms, nautical terms, animals, and instruments.[10] Duarte de Barbosa, the Portuguese traveller in the first decade of the sixteenth century had noted that the Tamil language spoken on the Coromandel Coast was different from the language that was spoken on the Malabar Coast. He noted that Tamil and Malayalam were very much like the Portuguese language generally spoken by the Castillians and others. Barbosa also thought that they were two different dialects of the same language.[11]

Portuguese-Tamil Dictionaries Since 1549

The mono-lingual dictionaries in Tamil available at that time were not useful to the Portuguese missionaries on a large scale. But they had a limited use for the missionaries who were not adequately skilled either in Tamil poetry or prose. Hence bi-lingual dictionary was the need of the hour with Portuguese meanings for Tamil words and Portuguese words for Tamil meanings. Therefore they planned to compose them during their course of missionary work in Tamil country. Anrique Anriquez the Jesuit who began to learn Tamil systematically desired to compose a *Vocabulario Luzitano em Tamulico* (Portuguese-Tamil dictionary), the new manual for scholars, introducing the new discipline of linguistics in Tamil that began in the sixteenth century. His first bilingual dictionary was prepared only by him without any assistance from others while working among the paravas on the pearl fishery coast. He began the work of the first bilingual dictionary as early as in 1548. In his letter dated 31 October 1548 to Ignatius of Loyola he stated that on arrival he began to learn Tamil and the language was difficult. He discovered therefore a sort of method of congregating words as in Latin. In a letter sent from the Jesuit College in Goa to Ignatius dated 9 September 1549 mention was made of Anriquez and his attempts made to compose a Portuguese-Tamil dictionary.[12]

After a period of three years we find again from his three letters dated 27 January 1552, 5 November 1552 and 6 November 1552 mention was made of still preparing the Portuguese-Tamil dictionary by Anriquez and the work in considerable progress. Again we find in the letter of 1558 that

Anriquez himself spoke of his Portuguese-Tamil vocabulary in the making for the use of the missionaries which he hoped to bring out soon.[13] The work was still in the making and in 1561 when he was in Nagapattinam he stated in his letter that Pero Luís the Tamil Brahmin convert was helping him in the work of dictionary.[14] Even after five years the work had continued to be in progress and we find mention of it as late as on 27 January 1566.[15] We do not know exactly when Anriquez had completed the work of Portuguese-Tamil dictionary.[16] Finally we hear the news in the Jesuit annual letter of 1601 sent to Rome while informing the death of Anrique Anriquez in 1600 that he had completed earlier the compilation of one Portuguese-Tamil dictionary. Scholars have mentioned that no extant copy of this first Portuguese-Tamil dictionary is available anywhere today. But I have found that it is preserved in the State Central Library of Goa.[17] A glimpse of the dictionary would be interesting to note the entry of words. There are 44 words under the letter entry of 'AB'.

ABC—ariviri, akshaviri
Abada animal v. unicornio—kaandaa mirugam
Abadivertidura—vadimbu, vizhlimbu
Abafadico —irukkamaanathilam, idanjal
Abafadamente—thittumuttu
Abainhar—Vudavatti thakkiradhu, porumiradhu, madiththu thaikiradhu
Abainhar espada—uraiyile podugiradu
Abaixar—pannikkradhu, thaazthigiradhu, thankkiradhu
Abaixar olhof—kan vizhillikradhu
Abalar v.boler, moverse— asaikiradhu, ullaikiradhu
Abalancar se moverse—keliiradhu
Abalizado h.v. afamado—peyarpondavan, kirthiudayavan
Abalo—alangal, asaivu, alaichal
Abanar v.avanar—veesukiradhu
Abarcar —adakki kozhlgiradhu
Abater—keezhlpaduthugiradhu
Abrigar v.amparar, abrigar-se—othungiradhu
Abrigo—othukkidam
Abril—chiththirai
Abrir—thirakkiradhu
Abrir os olhof—kan thirakkiradhu milikkiradhu
Abrir opeixe—keerugiradhu, pilakkiradhu
Abrir sea terra parede—vedikiradhu

Abrir se oflor—malarugiradhu, muzhilikiradhu, alarugiradhu, poopikkiradhu
Abrogar—azhlikkiradhu, thazhllukirathu
Abrochos—nerinji muzhllu
Abrochos de ferro—adi otti
Absolver—paava vimosan pannugiradhu
Absolivacao—paavavimosanam
Absolutamente—kevalamaay yeppadium
Absoluto—keval karthan
Absonavoz—inbam illaatha suram
Absorta—paava sammana ponavan
Abstero —mugam kudaathavan
Absterse—orrukiradhu, vilakiradhu
Abstinence no comer—ubavaasam, pattini
Absurdo—abaththam, abaththamaana
Abundar—thiralaai irukkiradhu
Abundancia—pooranam, sampooranam,
Abuzar—akkiramam pannugiradhu
Abuzo—thoorasaaram, anaasaaram
Abutica—ammukira
Abutre—kazhugu

The dictionary under alphabet 'X' contains only 5 words. There is no word given under 'Y' alphabet. The dictionary ends with 15 words in 'Z' alphabet.

Xadry—sathurangam
Xarela de coco—kottangkachchi
Xarope—kudineer, kashaayam
Xarrofa—nottakaaran
Xarenga—peechaang kuzhal

Zamboa—naarthangaay
Zambro—mooppadi kaalam
Zambolura—naaval maram
Zangao—kuzhlavi, vandu
Zanolha—vakkira kanni
Zangarrear—thaaru maaraay vaathiyaththai vaasikkiradhu
Zarogalhada—thinai
Zaravatana—nachchu kuzhlal
Zariaotinto—saathlingam

Zargunha—kuththukoll
Zebra—pettai koveru kazhuthai
Zenit—uchchi, uchcham
Zombar—mulamulavengiradhu, isaikiradhu
Zurraroasno—kazhuthai koopidugiradhu
Zurrir—kolaalay veesugiradhu, maarugiradhu

We have no recorded evidence as to suggest whether this dictionary of Anrique Anriquez was ever printed, while some of his other literary works dealing Catholicism were by that time had been printed and circulated to the Tamil world of audience.

Balthasar da Costa the Jesuit who had an excellent command over Tamil and Portuguese began to compile a *Vocabulario Luzitano em Tamulico* (Portuguese-Tamil Dictionary). His dictionary was in the making for many years as it had been reported in missionary records.[18] He completed it finally in 1660 and died in 1673. A manuscript copy of his Portuguese-Tamil Dictionary exists at the State Central Library in Goa.[19] The front page and the first page are not clear. A glimpse of the dictionary is interesting to note the entry of words. It runs as follows:

Abada animal v. unicornio—kaandaa mirugam
Abadivertidura—vadimbu, vizhlimbu
Abafadico—irukkamaanathilam, idanjal
Abafadamente—thittumuttu
Abainhar—Vudavatti thakkiradhu, porumiradhu, madiththu thaikiradhu
Abainhar espada—uraiyile podugiradu
Abaixar—pannikkradhu, thaazthigiradhu, thankkiradhu
Abaixar olhof—kan vizhillikradhu
Abalar v.boler, moverse— asaikiradhu, ullaikiradhu
Abalancar se moverse—keliiradhu
Abalizado h.v. afamado—peyarpondavan, kirthiudayavan
Abalo—alangal, asaivu, alaichal
Abanar v.avanar—veesukiradhu
Abarcar —adakki kozhlgiradhu
Abatar—keezhlpaduthugiradhu
Abrigar v.amparar, abrigar-se—othungiradhu
Abrigo—othukkidam
Abril—chiththirai
Abrir—thirakkiradhu
Abrir os olhof—kan thirakkiradhu milikkiradhu

Abrir opeixe—keerugiradhu, pilakkiradhu

Abrir sea terra parede—vedikiradhu

Abrir se oflor—malarugiradhu, muzhilikiradhu, alarugiradhu, poopikkiradhu

Abrogar—azhlikkiradhu, thazhllukirathu

Abrochos—nerinji muzhllu

Abrochos de ferro—adi otti

Absolver—paava vimosan pannugiradhu

Absolivacao—paavavimosanam

Absolutamente—kevalamaay yeppadium

Absoluto—keval karthan

Absonavoz—inbam illaatha suram

Absorta—paava sammana ponavan

Abstero —mugam kudaathavan

Absterse—orrukiradhu, vilakiradhu

Abstinence no comer—ubavaasam, pattini

Absurdo—abaththam, abaththamaana

Abundar—thiralaai irukkiradhu

Abundancia—pooranam, sampooranam,

Abuzar—akkiramam pannugiradhu

Abuzo—thoorasaaram, anaasaaram

Abutica—ammukira

Abutre—kazhugu

The dictionary under alphabet 'X' contains 7 words of entry. There is no word given under 'Y' alphabet. The dictionary ends with 21 words in 'Z' alphabet. Therefore certainly the dictionary of Balthasar da Costa is an improvement over the dictionary of Anrique Anriquez.

Xadres—sathurangam

Xaque nos xadres— sathuranga arasu

Xarela do coco—kottaivaachchi kottaivichchi, sirattai

Xarope—kudineer, kashaayam

Xarrofa—nottakaaran, vattakaaran

Xerenga—peechaang kuzhal

Xerengar—peechaang kuzhalaaley peechugiradhu

Zamboa—naarthangaay

Zambro—mudikaalan

Zambolura—naaval maram

Zangao—kuzhlavi, vandu

Zamolha—vakkira kanni, pottachchi
Zangarrear—thaaru maaraay vaakkyaththai vaasikkiradhu
Zarogalhada—thirina
Zaravatana—nachchu kuzhlal, sungaththan kuzhal
Zariao tinta—saathlingam
Zargunha—kuththukoll
Zavancria—vanda vellaatti
Zebra—pettai koveru kazhuthai
Zelozo—dharmampaththi, roshamullavan
Zelo—dharmamampaththi, rosham
Zenit—uchchi, uchcham
Zombaria—parigaasam, pariyaasam
Zorte —milechcha guru
Zumbar as abelhaf—mulamulavengiradhu, isaikiradhu
Zurraroasno—kazhuthai koopidugiradhu
Zurro—kazhuthai saththam
Zurrir—kolaalay veesugiradhu, maarugiradhu

The dictionary manuscript has 340 pages. It does not correspond to the manuscript of Portuguese-Tamil Dictionary preserved at the National Library in Paris.[20]

Nikolaus Dal (1690-1747) the Protestant missionary in Tranquebar had begun to prepare a vocabulary in Portuguese and Tamil in 1730. A fragment of this Portuguese-Tamil dictionary is preserved at the Archiv der Franckesche Stiftungen in Halle. The dictionary gives each word in Portuguese, Tamil and finally the pronunciation of Tamil in transliteration in German. The work begins with the word brinco/vilayattu/Weleiattu and it ends with the word finally with Cagar/belugiradhu/Pelughiradhu.[21] Nikolaus Dal had completed the dictionary work containing those nouns, verbs and adverbs of the two languages which are most used in daily interaction. It was published from the Tranquebar press on 22 September 1731. In this book we find Tamil words are printed in Tamil characters, juxtaposed with Portuguese letters for the use of the children in Portuguese school at Tranquebar.[22]

Onarato de Udine the Capuchin in Madras had compiled a Portuguese-Tamil dictionary in 1793. This manuscript is titled '*Vocabulario das palavars Malabares mais frequenters nos discursos, com a sua explicacao em idioma Portuguez*' and it is preserved in the Biblioteca Nazionale Centrale, Vittorio Emmanule II, Roma. The dictionary consists of five brief thematic Tamil-Portuguese word lists such as '*as partes do corpo humano, as paixoes do*

Homem, as tres potencias *da alma, Os cinco sentidos and do deo e dos elementos*'.[23] It is interesting to find that Onarato de Udine had put up three entries for the word rice i.e., arroz, arisi in Tamil; arroz cosido, soru in Tamil and finally arroz frio, neersoru/pazhaiyathu in his dictionary.

Tamil-Portuguese Dictionaries Since 1640

Ignacio Bruno (1585-1659) who lived and did missionary work for thirty years in Jaffna prepared a Tamil-Portuguese dictionary around 1640. According to Antão Proença (1625-1666) the Jesuit missionary, words mostly used in the poetical works in Tamil were abundant in the work of Bruno.[24] João de Sequeira (1673-1724) an Augustinian was in Mylapore and other places on the Tamil coast.[25] He began his dictionary work in 1716 and had compiled a *Vocabulario Tamulico em Luzitano* (Tamil-Portuguese dictionary) and completed it in 1718. A copy of this dictionary is preserved at the Albert Royal Library in Brussles in Belgium.[26] It has 262 folios in total with the dimensions of 13 x 18 cms. According to the note attached to the manuscript copy it is learnt that it was owned by Lord Stuart de Rothesay in 1855. The first page is in a different handwriting. This *Dictionario Tamulico-Grandonico seu potius malabarico grandonicum compositum opera et studio P. Joannis de Cerqueyra* on each page has Tamil word and equivalent Portuguese translation or description. In general we find that there are 30 to 40 words per page. Though the manuscript is fine it is rather in small handwriting. The dictionary runs as follows:

Aba—C. baldada
Abatooru—Infamia
Abgarikkiradhu—Furtar
Abgarippikkiradhu—Fazer furtar
Abagarippu—Furto
Abagarikka—Cousa furtada, ouque furtou
Abgariyaadha—Cousa na furtada, ou que nao furtoa
Abgarikkugai—O furtar
Abagadam—Engeno, q se Urde bo animo
Abagaaram—Disfavor
Abakeerthi—Infamia
Abaiyam—Seguraca se medo
Abayastham—Hum sinal, que os Reis faze co a mao untada co a safrao, em hum pano, e o dao por seguranca, do que o recebe ex
Abalam—Risco, a que hu sepoe. item, mal
Abam—Baldameto, ou cousa baldada

Abakkiyaathi—Injuria, que se dis salfamente se fundamento
Abayamaay—Baldadamente
Abakarikirigai—Exequias
Abaraatham—Penitencia, pena, ou multa, q se impoe; co-(vaangirathu) Apenar, multar, condenar-co– (kodukkirathu) Pagar a pena. A multa com. (kattaliidigirathu) Dar ou impor penitence. co. (thirkirathu) Coprir a penitencia
Abaaranam—Horas de uesporas
Abaththam—Despropofito
Abaththa visesham—Idem
Abavaatham—Descredito, defeza honra
Abasaanam—Descortesia q se fas
Abasaaram—Descortesia de palavra
Abaseyam—Rota na Guerra ou ser vebcido na Guerra, no jbgo xcz
Abaayam—Engano, ou mal que se fas; he o contrario, de (ubaayam)
Abaaya thanthiram, Abaayubaayam—Traca, ardilin malu
Abaaanatha vaaram—Traseiro, modeste
Abaakkiyam—Mosina desgraca
Abaakiyan—H. mosino
Abaakiyi—M. mosino
Abaavam—Nada, neggacao da realidade. gr
Abaiyam—Grito de quem pede justica.com. ritar, pedindo favor, ou justica com. – fazer justica &c
Abeshai—Desejo
Abeshikkira—C. que deseja, ou desejada
Abeshikkiradhu—Desejar
Abeshiyaatha— deseja ou nao desejada
Abimandirichcha—C. benta.
Abimandiriyaatha—C. nao beta
Abimandiripikkirradhu—Fazer benzer
Abimandirippu—Renzimento
Abimandirikkira—C. que actualmente benzem, ou esta benzendo
Abimanirikkiradhu—Benzer rezando
Abhimaanam—A quillo deque hum se preza, ou aprezar se de aluma C. con (Indha sasthirathile yenakku abimaanamundu) – Prezome desta Siecia. Apud Xutres sinifica as partes secretas, modeste
Abimanniyam—Pro.- na seunda significacao
Abin, Abin marundhu—Ansiao
Abippiraayam—Sentido, ou intento das palavras

Abirangak kaasu—Venezea no, moeda
Abisaari—Mulher adultera
Abishegam—Vneao
Abeesham—Desejo efficaz
Abhoorva— para exquirita
Abhoorvam—Idem, fubst
Abhogam—False, q se impoem sem fundamento

I have taken a sample of words under 'AB' only and the dictionary contains 55 entries. João de Sequeira the compiler had given the usage and the context and also referring it pronoun and other details and how it is to be used in grammar.

Domingos Madeyra (1685-1757), the Jesuit worked in the Madurai mission compiled a *Vocabulario-Thamulico-Lusitano* (Tamil-Portuguese dictionary). He says that this has been prepared for the use of the missionaries of the Society of Jesus. The dictionary begins with a dedication to Blessed Virgin Mary. It records that the work was completed in the year 1750.[27] We find the vocabulary beginning with the Tamil word A. The dictionary ends at page 165. There is one more additional page which contains Tamil alphabets at the end. Thus the dictionary has 166 pages in total.

The prologue of the text has three pages in the manuscript. Domingos Madeyra mentions that it was not his intention, to the gentle missionary reader, to prey on his wisdom or to offend the prudence of his judgment, in that the reader should refrain from critiquing this dictionary which is certainly be deserving of censure, for one always can find a sufficient reasons for our abilities not to be considered inferior or deserving of criticism. Domingos Madeyra the compiler mentions that he should like to clarify the motives which had led him to produce this somewhat limited work. For example, He was asked by various priests to include different things that were found missing in the earlier dictionary of the others. He opines that there was no dictionary for the common benefit of the beginners in Tamil language in the Jesuit Province of Madurai. Many dictionaries have been though composed by different authors in the more elevated style, apart from the astuteness and skill of the then recent one by Constanzo Giuseppe Beschi, which, without exaggeration, has no equal either in ordinary Tamil, which is called kodunthamizh or in elevated Tamil called senthamizh, used for poetry. The dictionary of Beschi is so fine that it exceeded in quality, even those of the native pundits, who rightly marvel at seeing such knowledge in a foreigner.

Domingos Madeyra had dared to undertake the compilation because it was a worthwhile exercise. Like a clever bee who gathers from the most fragrant flowers only what she needs Madeyra wished to produce the sweet honeycomb for the common-good. Hence he spared no effort in gathering together the many grammar rules and proofs of them, in order to assemble something like a tasty honeycomb. This was to facilitate and stimulate the reader who would study and use what is contained in this dictionary. He does not wish to induce boredom, nor so brief as to omit doubts about obscurities. Following the proverb 'Verbosity provokes aversion; Brevity invites approbation', he had, in effect adapted it in order to encourage everyone to learn its contents.[28]

In so far as concerns the Tamil grammar, he wrote nothing. Since his age of 65 had not allowed besides his poor health that prevented him from writing it all. He would take it that the grammar of Tamil had already been studied, so he will deal only with the words of the dictionary, including the verbs: to these, he assigned the past, future, affirmative, negative, imperative, infinitive and certain participles. These are the most necessary things needed for the beginners. Indeed, the determination of the past tenses of verbs is extremely troublesome in Tamil language, and only over a period of time they were learnt. He confessed that though he had served in the Madurai Mission for 30 years, he was still unsure of some of them, but after consulting certain experts, he could fulfill his promise.

By the grace of God, he hoped to satisfy all, like the other David, with the help of the words, like pebbles from the sling of eloquence, which guides them to the bulls' eye of the Goliath of Heathenism. Domingos Madeyra trusted that they achieve this, where others would surrender, saying 'Lord, What was that you want me to do?'. Putting themselves into the hands of the ministers of the vineyard of the Lord, through whose guidance they will be like, so that they would also be qualified to teach others what was necessary for their salvation.

At the end of the dictionary, Madeyra added a postscript, in which he put the major part of Beschi's own introduction, as he thought would help significantly in composition, reading and also writing, together with that of certain letters, in which cases they used single or double letters or when we changed them into others according to the rules of grammar but preserving the meaning.[29]

In the end of the manuscript besides finding the extracts from the preface to Beschi's '*Grammatica-Latino-Tamulica*' (printed at Tranquebar in 1738) a note in Portuguese is followed to the effect that, following the example of

Antão Proença in his '*Vocabulario Tamulico Lusitano*' printed at Ambalakaadu in 1679, where the Latin script order of entries had been preferred to that of the Tamil. Domingos Madeyra says that he does not use the order of the Tamil alphabet, because he judged it to be too confusing for the beginners. After hearing a word, to find a word in the dictionary one needs to read through the whole list of those beginning with the same letter. As they already knew the order of the Portuguese alphabet, they can easily search and find words very quickly in the dictionary.

Tamil-German Dictionary and Tamil-Latin Dictionary Compiled by Ziegenbalg, 1710-18

Bartholomäus Ziegenbalg stated in 1710 that he had begun compiling a dictionary and that in such a way as to write first of all, each word in Tamil character to which he immediately added Latin transcription to show how the word was to be pronounced correctly and after that gave the meaning in German. According to him one might wish that this Tamil language were taught and learned in Europe with as much diligence as other oriental languages, in as much as the Tamils were a great and innumerable people who might thus be helped out of their heathenish blindness by the grace of God. If all the protestant kings in Europe would assist in this and allocate sufficient funds, it might be also to their own advantage that one would be able to understand from their writings the profound mystery of their theology and philosophy in which one may find perhaps as much heathenish scribes.[30]

The missionary report of the year 1712 sent from Tranquebar mentioned that the Tamil-German dictionary had been completed and it comprised more than 40,000 words in such an arrangement the first came the preliminaries, and then the derivatives in addition a few phrases. As the dictionary work was begun in 1710 in which the letter 'A' completed had been sent to Europe. The dictionary work compilation consumed too much and paper and it had to been written subsequently on palm-leaves, otherwise, Ziegenbalg said that from all the Tamil works a dictionary of some 20,000 words that had even been previously collected, were to be written by one's own hand on paper. He described the procedure and method thus that the Tamil words written in its own characters, afterwards the way it was to be pronounced was given in Latin letters with transcription and at the same time thirdly he gave the meaning of the Tamil word in German along with it he also completed a German index.[31] Ziegenbalg also wrote that this dictionary was prepared for the use and benefit of the fellow Protestant missionaries who would follow to come to Tamil coast for missionary work. Although one did not wish to advise anyone that such a language can be learned merely from dictionaries, but on the contrary he said that it would be better to learn by reading the Tamil books written, from which missionaries would each time note down all the used and unknown words and committed them to memory.[32]

In 1716 Ziegenbalg also explained how he collected the words. He said that he not only read historical, theological and philosophical writings but also medical and philological texts in Tamil in order to find words for his dictionary. Of the poetical language Ziegenbalg said that there was as much difference between it and that employed in daily life, as between Latin and German. He collected moreover another 17, 000 poetical words for the dictionary out of all kinds of poetical writings, under the guidance of the experienced Tamil poets. This dictionary would help him but little, when he read and preached the Word of God, but it was the key to all poetical books in Tamil. In the catalogue of books titled '*Verlangtes verzeichnib unserver Malabarischen Bucher*', it has been mentioned again and again that Ziegenbalg had copied many beautiful phrases and words from Tamil texts such as Bhaarta Urai, Arichandran Kathai, Thiruvilayadal Puranam and Thamizh Arivaazhl.[33] Thus he had also taken great trouble to understand the worship of the Hindus from their own books.

His dictionary had therefore in total had grown to around 40,000 words and expressions and he had arranged according to the Tamil alphabet so that the derived and compound words might be found near their source. Ziegenbalg considered his dictionary as a great treasure not so much on account of the great care and trouble that he had spent on them, as in consideration of the uses to which they might be put. Ziegenbalg's Tamil-German dictionary manuscript is no where traceable in the repositories. The detailed reports sent from Tranquebar between 1706 and 1718 furnish the details.[34]

However, a Tamil-Latin Dictionary was also in the making by Ziegenbalg. A half sheet had been printed in Tranquebar in 1719. It is used as a waste for binding the printed Tamil Bible in 1720 and we find it in the book preserved in the National Library, Paris.[35]

Latin-Tamil Dictionary of Bourzés (1710) and his Tamil-Latin Dictionary (1731)

Louis-Noel de Bourzés the Jesuit was stationed at Madurai in the beginning and he later became the provincial of the Madurai Mission in 1713.[36] He began compiling a Latin-Tamil dictionary (*Dictionarium Latino-Tamulicum*) in 1708 completed in 1710, as noted in Jesuit records, of which the manuscript could not be traced any where now. While staying on the pearl fishery coast he had used it for his own purpose.[37]

Bourzés had also begun to compile a Tamil-Latin dictionary (*Dictionarium Tamulico-Latinum*) and completed the work in 1731 as found mention in the Jesuit correspondences.[38] This manuscript is also not traceable anywhere now but Constanzo Giuseppe Beschi said that he had used it.[39]

Tamil-Latin Dictionary of Beschi, 1742

Beschi decided to compose *Vulgaris Tamulicae Linguae Dictionarium Tamulico-Latinum* (Tamil-Latin Dictionary of the ordinary Tamil) to which are prefaced several rules necessary to be read beforehand. This *Dictionarium Tamulico-Latinum* work he began in 1741 and completed it in 1742.[40] The State Central Library in Goa has a copy of the Tamil Latin dictionary and it is mentioned clearly that the copy was made in 1744.[41] In the preface (vide note number three) Beschi made it clear that the words from other languages that came to be used in Tamil were marked with special signs. He stated that the words found in this dictionary have been mainly from the mission area of Madurai. There are around 9000 words in this dictionary.

Beschi mentions many words which he had taken from the works of Roberto Nobili for his Tamil-Latin dictionary making with the mention 'apud R.P. Robertus'. Some of these words include yettrinam, krithikiradhu, samoodhidham, sannidhaanam, saanithiyam, sudhan, suvikaram/suvikaaram, suththiram, somam, gnaandhikkiyar, thama, dhikkaaram, devokiththam, naathakirthiyar, paraathasangai, priyam, pradhamam, vikkirapusagun, viseshanam, vilaasam, vinaapudham, vedhapaaragar, sthaanaathipathi and snaabhagun.

I have selected from this Tamil-Latin dictionary with words under 'a' because it is interesting to note for Beschi had given the various meanings and the context.

Agosarum—quod inaccessible est seu visu seu dictum

Akkiyaanam—inscientia, ignorantia ex

Akkiyani—ignorans, vulgo

Anaadhi—qui est sine principio

Anaachaaram—inurbaniitus

Anukarana kaaranam—causa concomitans

Anumaanam—signum unde; Devadiraviya anumaanam signum divitarum divinarum

Anrupam—conformitas, rupam, figura anu, simul
Angigaaram—assensus interior
Angikarikkiradhu—assentiri in animo
Anthakkaranam—interiors animae facultatres
Appiyamusiththam—est sacrificium ad expianda quinque criminal in usu plantarum
Asakasaanthiram—asam ovis kasam elephas et
Asattai—negligentia praecipue ex eo quod vili pendat alterius dicta
Asareeri—corporeus et
Asanam—1. cibus 2. comestio
Athithoothan—praecipuus legatus
Arcsiyashishta—sanctus
Archiyam—veneration
Avasthai—1. Necessitas 2. Vulgo articulus mortis; Avsathai pusuthal sacramentum extremae unctionis.3. finis Unde addito privatvo fit

Thus Beschi had given the word 'avasthai' with three different meanings. Similarly 'sahasam' in his Tamil-Latin dictionary has been given with three different meanings in use such as (1) justitia originalis (2) dicere sincero prout est (3) veluti cujuslibet natura referring to original justice; to say a thing as it is; and as some one's nature. Similarly he gave the word 'anumaanam' with two different meanings in which it has been used by Tamils such as sacrament a sign of divine riches and signs that are doubtful.

Beschi had given several terms related to various sickness and diseases in this Tamil-Latin dictionary. We find Tamil terms such as Akkaram, iraippu, kadukkirathu, karappan, kaaychal, vidungkaaychal, vidaakaaychal, Naalamurai, munraamuari, kulirkaaychal, surakaaychal, varaikaaychal, utkaaychal, pithakaaychal, sileshamkaaychal, vaathakaaychal, manthakaaychal, kadungkaaychal, doshakaaychal, kushtam, pilavai and kiraani neerazhivu.

There are also words used and found from the areas like Pondicherry in the dictionary and we find entries. Examples include, Garudan-Falco Pondicherianus; Puthucherry-Pondichery; Puduvai-vid Puthucherry; Vaangu1.q.Pugio 2. pro.visipalagai; Vaasagappa-vid. Vaasappu; Vaasaappu-tragaedia, comedia. It may be stated that Pondicherry was then within the jurisdiction of the Madurai province Jesuits. In the end of the dictionary the word beginning from skandam till srimuga at the end we find 94 Sanskrit words given in Tamil and their meanings found in Latin. The dictionary ends with the phrase 'Finis Laus Deo et B.M.V'.

I have gone through the manuscripts preserved in London and Copenhagen.[42] I have also examined the manuscript copies at Shenbaganur and Pondicherry.[43] An analysis and a comparative study of them shows no variance. There are seventeen paragraphs found in the preface of Beschi in the manuscript. Under the caption 'year' Beschi had given 60 years of Tamil calendar, with the example beginning with prabhava 1747-1748 and till akshaya 1806-1807. Since he mentions particularly against the year 1742, by referring it to this year (which corresponds to Dhundumi, the Tamil year), we may infer that he had completed the work of dictionary in 1742. Beschi had cited kural no.1 (Agara muthala vezhuthellaam) and no. 127 (Yaagaavaraayinu Naakaaka) in his Tamil–Latin Dictionary.

Portuguese-Latin-Tamil Dictionary of Beschi, 1743

Beschi began his another work 'Second part of the dictionary of the ordinary Tamil language, in which words of the ordinary Tamil language are added to Portuguese words explained in Latin'.[44] This Portuguese-Latin-Tamil dictionary work he commenced in 1742 and completed it in 1743. The State Central Library in Goa has a copy of the Latin-Tamil dictionary where exclusive mention has been made that the copy manuscript was made in 1744.[45] There is also a manuscript exists at the Madurai Province Jesuit Archives, Shenbaganur. Out of the two manuscripts of Portuguese-Latin-Tamil dictionary, kept there (Under MSS Tam, no.3 and MSS Tam, no. 5) one is appended to the Chathuragaarathi of Beschi and it contains 103 pages. Each word is given a number and it has total of 4073 words. In a separate page in the beginning of the dictionary we read as follows: '*A.M.D.G; Vulgaris Tamulicae Linguae Dictionarium Pars Secunda Mque Lusitanis Vocibus Lattine Explicatis Vulgaris Idiomatis Tamulicae Verba Ajunguntur Auctore P. Constantino Josepho Beschio Soc. Jesu Missionario*'. There is a preface in Latin followed by the above note.

An analysis of the Portuguese-Latin-Tamil-Dictionary appended along with Chaturagaraathi manuscript with the other independent Portuguese-Latin-Tamil dictionary manuscript preserved at the Madurai Province Jesuit Archives at Shenbaganur we find interesting and revealing information. The independent manuscript copy has no first page and the last page. There is a note and it states that there is an omission of 44 words in the first page and 29 words in the last page. This dictionary has meanings for 4353 words. The manuscript copy contains numerals for each word and so accordingly the page two starts with 45th word and hence 44 words omitted had been said to be correct. The last word numeral is 4073 and so the rest 29 words said to be found missing at the last page must bring a total of 4102 words.

Beschi's Portuguese-Latin-Tamil dictionary copy made later by scribes is separately kept in the Madurai Province Jesuit Archives where the first and last the page is found missing. But there is a note recording that forty-four words in the first page and twenty-nine words at the last page are found missing. The dictionary contains 4353 Portuguese words and meanings as noted by Leon Besse (1853-1919). This shows that new words have been added to this copy of the dictionary and the work undertaken was in progress by some one else.

In the preface of the dictionary Beschi writes that, since almost all the people in India knew Portuguese words, he had given them at first. However he does not deny the fact that many people are unable to understand these words. Therefore, to remedy the defect, he had added (along with Portuguese words) also the Latin words. Thus, Portuguese words alone are found in alphabetical order and all their meanings can be found in Latin. In order to obtain the fruits of the two, he could not avoid this work of doubling. As he chose the words as minimum as possible, this work was not a large one. Those words in Portuguese which had multiple meanings had not been given. Those words which were frequently used and the very significant ones alone had been provided. Ordinarily, verbs and nouns are given either before or after. The adjectives and adverbs are separately given either before or after followed by the respective words. Hence, the dictionary in his opinion fulfilled the purpose all that what was in the mind of a person can be understood easily by the other. So also, all the Portuguese words and their corresponding Tamil words are not given. But those words which are very frequently used alone are provided. The structure of the verbs and the conjugations are not given because all the people knew it from his Tamil-Latin dictionary. The aim of Portuguese-Latin-Tamil dictionary therefore was to enable to help those children who would begin to speak with common words and to proceed to pave way for the difficult words. As a precaution, he had removed those awkward words from the Tamil-Latin dictionary. This also he had done with a view that it was necessary for a learned man, because we must understand, while others express it. Thus, for this brevity, he was sure that no one would find fault with him. Those new-comers should read cautiously those that he had said very carefully and keep it in their mind and should use this dictionary. The users will feel that there is something in this dictionary to make them interested.[46] Thus Portuguese words had been given at first in the alphabetical order in the dictionary and it has been prepared for those who do not know Tamil. In this work Beschi therefore followed the Portuguese words in alphabetical order only. Thus words first come such as aba, abe, abo, abr, abs, abu, ar, at, ax and az.

All Portuguese words in the dictionary did not have the corresponding Latin meanings too. He then mentioned the declension of nouns cited in his *Grammatica LatinoTamulicaa ubi de vulgari Tamulicae linguae idiomate kodunthamizh dicto ad Usum Missionariorum Soc. Iesu* (Tranquebar, Typis Missionis Danicae, 1738).[47] Then he continued to explain the conjugation of verbs and the rules of the grammar. He had thus taken adequate care to see that there was the least deviation from the rules.[48]

There is a complete difference between the Portuguese-LatinTamil dictionary of Beschi and the other dictionaries composed by him. It is to be noted that the Tamil words first appear in alphabetical order in the Chathuragaraathi and Tamil-Latin dictionary, whereas in the Portuguese-Latin-Tamil dictionary we find the words in Portuguese alphabetical order. Thus first comes, words with beginning such as Aba, Abe, Abo, Abr, Abs, Abu, Ar, At, Ax, Az. Thus words of Ac, Ad, Ag, Ah, Ai, Am, An, Ao, Ap, Ar, At, Ax, and Az had been appearing and it followed by words of Ba, Be, etc., The dictionary finally ends with word of Z. It is found that all Portuguese words did not have the corresponding Latin meanings vide MSS folio 4 and therefore some Portuguese words had only Tamil corresponding words. A closer examination of the Portuguese-Latin-Tamil dictionary reveals that some meanings are given by making sentences vide MSS folio 3, such as '*Licet aspexerim, examina verim omnia, non inveni sic composia*' '*Accidit it puncto non connigint in an*' as earlier done by Beschi in his Tamil-Latin dictionary. Sometimes Thirukural is also used for explaining meanings in sentences vide MSS folio seven such as Anbukkum undo adaikum Thazhl, in Valluvar's chapter titled Anbudamai, kural no.1. Similarly medical terms are also found at folio 8. Active voice and passive voice had also been given at folio two. Beschi's Portuguese-Latin-Tamil dictionary gives all the words either decent or less decent and phrases and proverbs as found in the ordinary life of the people which the missionaries will have to hear and understand. This dictionary served to help to find the appropriate Tamil expressions. Beschi in this work had confined himself to just those items of immediate use to missionaries. To cite an example he has mentioned only two varieties of rice such as arros crus (raw rice) and arros cosido (cooked rice). The Shenbaganur copy of Portuguese-Latin-Tamil dictionary was used by Fr Ford, a Jesuit of the Madurai Mission who prepared a French-Tamil dictionary (Petit Dictionnaire Français-Tamoul) and published by the South India Times Press at Tiruchirapalli in 1881.

A copy of the Portuguese-Latin-Tamil dictionary of Beschi is preserved at the British Library, London, titled *Vulgaris Tamulicae Linguae Dictionarium, Para Segunda, in qua Lusitanis Vocibus Latinae Explicatis*

Vulgaris Idiomatis Tamulica Verba Adjunguntar, under Additional Manuscripts, No. 26110. It has the dimensions of 17.7 x 22.2 cms with a total of 313 folios. The first part of the work contains Tamil-Latin dictionary and the second part was Portuguese-Latin-Tamil dictionary with three appendices with a total of 311 sheets. At the end of the manuscript it is recorded as follows: 'This transcription of the dictionary was completed by me, Paschoal Manoel de Reys, on Thursday noon of 26 February 1750'. Another copy of the same Portuguese-Latin-Tamil dictionary of Beschi is available at John Rylands University, Manchester under MS no. 11. It has the dimensions of 30 x 24 cms with a total of 114 folios.

It may be pointed out that Beschi's Portuguese-Latin-Tamil dictionary manuscript and Proença's Tamil-Portuguese dictionary printed in 1679 had been different in style and method. Proença's dictionary gives Tamil words first and gives Portuguese meanings. Beschi gives Portuguese words first and next the Tamil meanings. Proença does not make Tamil words in Tamil alphabetical order and also gives Portuguese alphabetical order. Beschi systematically follows the Portuguese word order. All Portuguese words are not given equivalent Latin words by Beschi. (Shenbaganur MS folio 4). Beschi uses Thirukural for some sentence expressions, such as anbukkum undo adaikum thazhl (folio 7).

Portuguese-Tamil-Singalese Dictionary of Jácome Gonçalves (1735) and Euzebio do Rosario (1772)

Jacome Goncalves born in Goa in 1676 joined the religious order of the Oratorians of Goa and was ordained a priest and sailed for Sri Lanka to do missionary work in 1706. He became the superior of the mission and later served as Vicar-General in Sri Lanka from 1724, and died on 17 July 1742. He had composed Portuguese-Tamil-Singalese dictionary in 1735.[49] The folios one and two contain Tamil scripts. Folios three and four contain Singala alphabets. The preface begins at folio 5 and ends at 5v. The grammar of Singala language composed by Jacome Goncalves is found from folio 6 to 11. The dictionary thereafter actually begins at folio 12 with the Portuguese word ABC and mentions in Tamil Arivari/Akshaavari and followed by the meaning in Singala finally. The dictionary ends at folio 277 with the word Zurir in Portuguese giving the Tamil meaning kolaalaye vissugiradhu/maarugiradhu. There is an appendix with the terms of spirituality, celestial bodies, seasons, months, days, numbers and parts of the body, ending with the folio at 291a. There is a copy of the same manuscript preserved at the Overseas Historical Archives in Lisbon.[50]

Euzebio do Rosario born in Goa joined the religious order of the Oratorians of Goa and he worked in Sri Lanka as a missionary. He had composed a vocabulary prescribed for those working at Tamil and Sinhalese, beginning with Portuguese, followed by Tamil and ending with Singalese in 1772.[51] It begins with ABC and gives meaning in Tamil Arivari/Akshaavari then gives in Singhalese. The dictionary ends with the Portuguese word Zurir giving the meaning in Tamil visugiradhu/maarugiradhu with Singalese word then at folio 162v. I have the omitted the singalese expressions as I am not familiar with. Given below are some of the terms which are not found in the dictionary of Jacome Goncalves.

Abade —Maamuni rikiya seshan
Abecedarrio—ariviri, akshaviri
Abegao—pannaikaaran
Abelha— theni
Abelhinha, crianca—theni kunju
Abelhao —madhu, kaama
Abencocar—aseervadhikiradhu vaazthugiradhu, mangalamaay sollugiradhu
Aberta abertura v.fondo, gieta—thirantha thirappu, thiravu, vedippu
Abespa—kuzhavi
Abespao —pey perung kuzhavi
Abilitar —pirupathi kudukkiradhu
Abetumar—kisir musugiradhu
Abeturno—anaadhi
Abismo—paathaazhlam
Abujrar—aanaei ittu saathikkiradhu
Abobada—thaaku, thaangu
Abobara—suraikkaay
Abobara de Guinea—parangikaay, sakkarai poosanikaay
Abobara de Goa—neersuraikaay
Abobreira—surai kodi
Abocanhar—thendi kavvi thinnugiradhu
Aboleriha—pori
Abominacao—veruppu, aruveruppu
Abominavel—verukkapadathakka

I have compared the contents of Portuguese-Tamil-Singalese Dictionary of Jacome Goncalaves of 1735 and that Euzebio do Rosario's dictionary. They do not match and tally in total. They are completely different from each other.

Tamil-Latin Dictionary of Johann Philipp Fabricius, 1772-3

Johann Philipp Fabricius made an effort to compile a Tamil-Latin dictionary in 1772. He carried out the work till February 1773. This manuscript is incomplete and it is preserved at the Archives of the Francke foundation in Halle.[52]

Realizing the Importance of Dictionary Paper Manuscripts and the Wide Usage

Balthasar Estevas da Cruz was an errand boy of Manoel de Jesus (1710-1800) the Augustinian Bishop of Mylapore Diocese from 13 April 1786 to 13 January 1800. Estevas da Cruz completed copying the manuscript of *Vocabulario Luysitano Tamulico* (Portuguese-Tamil Dictionary) on 19 September 1738.[53] The original author is not stated and the manuscript has 480 folios. The dictionary begins with letter ABC in Portuguese giving the Tamil meaning Ariviri/Akshavari, and ends with the word Zurzir in Portuguese giving the meaning in Tamil visugiradhu/maarugiradhu. There is a note attached to this manuscript wherein it has been stated that this copy of Portuguese-Tamil dictionary belonged to Francisco do Carmo the Franciscan missionary who used it during 1809-1817.[54]

Many Catholic missionaries in the eighteenth century had continued to use the dictionaries prepared by the earlier missionaries. There is a copy of Tamil-Portuguese dictionary which belonged to Francisco de Noronha, stationed at Pesalai, in Thalai Mannar, of the Congregation of the Oratorians of Goa the missionary in Sri Lanka dated 8 November 1838.[55] This dictionary contains 60 Tamil year cycle from prabhava-vibhava (in 1627- 1628) to krodhana/atchaya (in 1686-1687) vide folios 117v-118. Further, subhanu year (1703-04), subagirithu year (1722-23) and krodi year (1724-25) had been mentioned vide folios 89v and 141. Therefore it is concluded that the original dictionary was of the seventeenth century and it had been also referring to the corresponding years when copies were made.

Similarly Gaspar de Santa Izabel who was born in 1770 at Panjim Goa joined the Augustinian religious order on 27 October 1783 and professed his vows in March 1786 came to Mylapore in 1798 to do missionary work and he was the vicar of the church at St. Thomas Mount and became the Vicar-General of the Mylapore diocese in 1806 and later died in Madras on 3 October 1818 and buried in the chapel of St. Rita in Mylapore. It has been reported that he used a Portuguese-Tamil dictionary wherein mention is made at the end of the manuscript (the original author name is unknown) and it is preserved now at the State Central Library in Goa.[56] This dictionary begins with the word Aba do vestido giving the Tamil meaning vadimbu/

vizhlimbu and gives the meaning in Tamil and ends with the word Zurir giving the meaning in Tamil visugiradhu/maarugiradhu. From the Tamil scripts used it is evident that the original dictionary was certainly of the seventeenth century.

Dictionaries were continued to be copied by scribes and these were widely circulated for use by the Catholic missionaries even in the overseas. A copy of the Portuguese-Tamil-Singhalese dictionary is now preserved at the library of the Geographical Society of Lisbon, Portugal.[57] At the end of dictionary manuscript in folio 163, wherein it has been recorded 'copied from a vocabulary of the venerable reverend father Euzebio do Rosario, the Catholic missionary'. The work was undertaken according to a request for use of this dictionary in St. Kitts Islands, on Monday, 9 July 1838. We know that this dictionary was prepared by Euzebio do Rosario in 1772. Thus it had been found that the dictionaries compiled by various missionaries were still more valuable, since many of them could not see the light of printing. Having sent the Tamil dictionary manuscripts with the request to preserve it in Europe and else where, we are able to get at least a glimpse of the dictionaries for research today.

There are also two undated and anonymous Portuguese-Tamil dictionaries available in the State Central Library in Goa but the calligraphy suggests that they belong to the seventeenth century.[58] There is one undated and anonymous Tamil-Portuguese dictionary too and the calligraphy suggests to the period of eighteenth century.[59]

It may be mentioned here that there are some missionaries who made efforts in dictionary related works. Christoph Theodosius Walther the missionary in Tranquebar made a study of Tamil-Arabic phrases in 1727. He also made a study of Hebrew-Tamil synonymy in 1740. These two manuscripts are preserved at the Royal Library Copenhagen.[60]

Further, a Danish-Portuguese-Tamil dictionary was prepared at Tranquebar the Danish settlement on the Tamil coast but the author and the year of completion is unknown. This manuscript has 97 folios and it is preserved in the Royal Library, Copenhagen.[61] It is possible that those Danish officials/servants of the company or the missionaries who lived at Tranquebar must have had attempted to compile such a work.

Dictionaries Prepared in Palm-Leaves and Sent from Tranquebar to Copenhegan, 1731-2

It is interesting to find the dictionary works in Tamil by the Protestant missionaries of Tranquebar made in palm-leaves. These were reported in a correspondence sent to Copenhagen in 1731. There are five such works that include (a) Agaraadhi peyar suvadiyin muthalaam vaguppu (A Tamil lexicon arranged in alphabetical order: Part one) partly revised by Zieganbalg sent to Madras in 1732 (b) Agaraadhi peyar suvadiyin irandaam vaguppu (A Tamil lexicon arranged in alphabetical order: Part two) partly revised by Zieganbalg sent to Madras in 1732, (c) Lusitam-peyar suvadiyinpadi yezhuthapatta muthalaam per chuvadi (The lexicon prepared based on a Portuguese dictionary: Part one) obtained from Jaffnapattanam which was still in the possession of the library in Tranquebar (d) Lusitanu-per suvadiyinpadi yezhuthapatta irrandam per chuvadi (The lexicon prepared based on a Portuguese dictionary: Part two) with alphabets from a to ve and (e) Pallikkudathin pillaigalukku padippikkira per chuvadi (Dictionary for school children). The Tamil schoolmaster of Tranquebar gave this name to this particular dictionary and the reason for its composition could no longer be found out since the author had passed away.[62]

New ideas for preparing various types of dictionaries had emerged based on the language dictionaries. C.T. Walther, the Protestant missionary in Tranquebar had also catalogued Tamil plants and compiled a botanical dictionary of Tamil plant names with description in Latin during 1732-39. The manuscript is preserved in the Francke Foundation at Halle in Germany.[63] Some Catholic missionaries like Francisco de Noronha Paulo (born in Goa in 1780 joined the Oratorian religious order came to Mylapore learnt Tamil and then and went to Mannar as a missionary and died in 1848) had prepared a dictionary for the express purpose of spirituality but indirectly serving the purpose of making Tamil dictionary. He in his work gave the meanings of all the words in the confession, for the greater convenience of those who were beginning to learn the Tamil language. This manuscript copy is preserved at the State Central Library in Goa.[64]

Missionaries established contact with people in various areas of the interior hinterland of Tamil country in the seventeenth century previously unknown moving away from the coastal areas. At that time they listened carefully to speakers of Tamil and they analyzed the unknown words and sentences, and they converted the spoken language into some kind of written form. This aspect of missionary work ultimately resulted in a huge body of linguistic documentation of Tamil found in dictionaries, vocabularies and teaching manuals.

To conclude it may be said that Anrique Anriquez, Balthasar da Costa and Antão Proença the three Portuguese missionaries who wrote Tamil grammar in Portuguese had compiled dictionaries. A general knowledge of

Tamil grammar was part of their skill that had enormously helped in making dictionaries. It is known that there was no clear distinction between vocabulary arranged alphabetically with definitions of the words and the dictionary arranged alphabetically words of language at that time. We find the Tamil lexical works of the natives had some impact, because a few missionaries like Louis-Noel de Bourzés and Constanzo Giuseppe Beschi studied them and cited in their own dictionaries compiled. It may be surmised that though the native lexical works were in poetic form, the missionaries had studied them extensively, examined and carefully used them in their venture of making bilingual dictionaries. The language of Tamil was very difficult to learn; and it was no easy task for the early missionaries to construct vocabularies. Some missionaries in the Tamil country had occupied themselves and wrote bilingual dictionaries. A few missionaries had attempted to compose even multilingual dictionaries or vocabularies or glossaries for their use. The purpose and scope of the dictionary, methods or the basic characters of such work, the plan and arrangement, and the sources of the dictionary had varied from individuals to individuals. Portuguese, Latin and French words borrowed into Tamil, treatment of philology, morphology, etymology of vocabulary, the problems confronted by them in the process had been recorded in the preface to the dictionaries.

An examination of the various manuscript dictionaries leads us to conclude that the dictionaries were compiled in the following manner. First of all, a list of Portuguese, Latin, German, French and English words was written on loose leaves (only the leaves were numbered not the page). Their Tamil equivalents were then added to the right of each Portuguese, Latin, German, French and English word, as and when they became available as can be deduced from the differences in the darkness and lightness of the ink strokes in the Tamil entries and also from the fact that some European language words were left with no European equivalents. It would appear to us that someone with a wide vocabulary supplied some of these entries. The crudeness of the Tamil hand its comparative immaturity and the spacing of the words suggest to us that the entries were written by the European missionaries, who presumably wrote what they heard and who added dictionary items of vulgar Tamil speech alongside the more vocabulary. A manuscript preserved at the house of Ananda Ranga Pillai gave me additional insights into the way how dictionary was in the making. Yet another manuscript of dictionary in making preserved in Paris also gave me the various steps involved. Though in the Paris copy it is stated the Tamil entries were copied by one person by name Raththinam in 1848 and the Portuguese entries were copied by another person in 1849, Tamil and Portuguese words are given only till page number 11. Thereafter only Tamil words are written and the Portuguese meanings remain incomplete. It had been stated that the manuscript was found in the family of Thiruvengadam Pillai in original having 290 pages. This was checked and collated and finally sent to France by Edouard Ariel in September 1849.[65]

Europeans have found Tamil nikandus as fine example of compilation of cognate words. They refer to many words that give to a single meaning but do not point out under what contexts they should be used. Europeans studied Tamil nikandus such as Pingala nikandu where they found under the head–solluthal, where they found that there were twenty two words given, but the places to be used were not given.[66] Hence they needed to go to the literary context under which the words used in order to know the meanings. Constanzo Giuseppe Beschi deserves a special place of mention because he had composed five dictionaries, such as Chathuragaraathi, Tamil-Latin dictionary, Portuguese-Latin-Tamil dictionary, Tamil-French dictionary and French-Tamil dictionary. No one else has compiled such great works in Tamil. Nikandus were in poetry were known to Beschi and so to arrange the words in alphabetical order, he gave meanings and laid foundation for the modern dictionaries. It must be said that Beschi had also composed the regional dialectical dictionary in Tamil.

Notes

1. *Tholkaapiyam*, South India Saiva Siddhanta Pathippu Kazhagam, Madras, 1982, Uri-iyal, 796.
2. Ibid., Idai-iyal, 750.
3. Ibid., Yechaviyal, sutra, 928-30.
4. Sundara Shanmuganar, *Tamil Akarathi Kalai*, Pondicherry, 1965; A. Dhamotharn, *Tamil Dictionaries-A Bibliography*, Wiesbaden, 1978; V. Jayadevan, *Tamil Akarathiiyal Valarchi Varalaru*, Madras, 1985; V. Veerasami, *Akarathikalai*, Madras, 1989.
5. Thandavaraya Mudaliar, ed., *Dhivaagaram*, Madras, 1835.
6. Gregory James, *Tamil Lexicography*, Tubingen, 1991.
7. Albertine Gaur, 'Bartholomäus Ziegenbalg's Verzeichnis der Malabarischen Bucher', *Journal of the Royal Asiatic Society of Great Britain and Ireland*, no. 3-4, October, 1967, pp. 63-95, see, p. 69.
8. Ibid., p. 68.
9. Narayana Ayyangar, *Puliyur Revana Siddhar Iyatrriya Agardhi Nikandu*, Madurai, Senthamizh Press, 1921. Revana Siddhar in his paayiram mentioned '*Ariya sagatham ayiratainutturu; Oru pathaarena uraithidum andinil...agarathi nigandena; Odinan yaavarum unarthida ninanthey*'. Sundra Shanmuganar, Tamil Akarathi Kalai, p. 38.

10. Biblioteca Municipal de Porto, MS 804; Elena Losada Soler, 'The Encounter of Languages. Reflection on the Language of the Other in Roteiro da Primeiro Viagem de Vasco da Gama' in Anthony Disney and Emily Booth, *Vasco da Gama and the Linking of Europe and Asia*, New Delhi, 2000, pp. 200-19.
11. Duarte Barbosa, *The Book of Duarte Barbosa: An Account of the Countries Bordering of the Indian Ocean and their inhabitants*, 2 vols., ed., L.M. Dames, London, 1918–21, rpt., Delhi, 1989, vol. I, pp. 84, 340; vol. II, pp. 71-73.
12. Raphael Ricci, *Selectae Indicarum Epistolae Nunc Primum Editae*, Florentiae, 1887, p. 120; Jose Wicki, *Documenta Indica* (hereafter DI), 18 vols., Roma, 1948-1988, vol. I, p. 492.
13. António da Silva Rego, *Documentação para a Historia das Missões do Padroado Portugues do Oriente*, 12 vols., Lisboã, 1947-58, vol. VII, p. 204, vol. V, p. 153.
14. Archivum Romanum Societatis Iesu (hereafter ARSI), Roma, MS Goa, 8 II, fls.336r-39v, see the letter of Anrique Anriquez to Laynez from Mannar, dated 29–30 December 1562.
15. DI, vol. II, p. 305, see the letter of 27 January 1552; António da Silva Rego, Documentação para a Historia das Missões vol. I, pp. 95, 153 see the letter of 6 November 1552; DI, vol. II, p. 751, see the letter of 27 January 1556; D.I. II, p. 492, see the letter of 1549; DI, vol. X , p. 159, see the letter of 1558; DI, vol. XI, document no.64, pp.16, 64-6, 750-1. The text runs as follows: '*Tinhamos a dias hum vocabulario feito. Este anno passando se trabahou de se acrecentar, e passarao de dois mil vocabulos que sao acrecentados*'.
16. DI, vol. VIII, p.751, see the letter of 27 January 1566; vol. X, p.159, see the letter of 1558.
17. State Central Library (hereafter SCL), Goa, MS 35, Pe. Anrique Anriquez, Vocabulario Lusitano-Tamulico Seguido d'uma Arte Tamulica, Século XVI; See also, Index materiam dequibus tracarunt Lusitane Provinciae scriptores ab initio Societatis ad annum 1724 in Antonio Franco, *Synopsis Annualium Soceitatis Jesu in Lusitana ab annis 1540 usque ad annum 1725*, Haeredum, MDLCCXXVI, p. 520.
18. António da Silva Rego, Documentação para a Historia das Missões, vol. VIII, pp.120-3; vol. IX, pp. 91-124, 135-138, 254, 471.
19. SCL, Goa, MS 49. Balthasar da Costa's work comprises of two parts, a dictionary and Portuguese-Tamil Grammar.
20. Bibliothèque Nationale, Paris (hereafter BNP), MSS Indien, no. 223.
21. Archiv der Franckesche Stiftungen (hereafter AFSt), Halle, M2 D4:30, fls.1-9.
22. Det Kongelige Bibliotek (hereafter KB), Kobenhavn, 47-459; KB, Hjelmst, 2037; Nikolaus Dal, *Quarta parte da grammatica Portugueza, convem a saber hum vocabulario em Portuguez e Malabar, que contem os nomes, verbos e adverbios de ambas as lingoas, que sao mais usados no trata cotidiano, e dos quaes os Vocabulos Malabares vao impressos com o carcater tamul, e juntamente com letra Portugueza para o uso da escola Portugueza de Trangambar*, Trangambar, Na Officina da Real Missao de Dinmarca, Anno de 1731. The same vocabulary containing Tamil printing is placed under Sanskrit. BNP, Collection of Eugene Burnouf, Sanscrit 1096. The pages 49 and 50 are missing and the last leaf is

mutilated. See also, *Der Königl. Dänischen Missionarien aus Ost-Indien eingesandter ausführlichen Berichten, Von dem Werck ihrs Ams unter den Heyden, angerichteten Schulen und Gemeinen, ereigneten Hindernissen und schweren Umstanden; Beschaffenheit des Malabarischen Heydenthums, gepflogenen brieflicher Correspondentz und mundlchen Unterredungen mit selbigen heyden*, (hereafter HB), Teil 1–9, (Continuationen 1–108) Waiserihaus, Halle, 1710–1772, Continuation 32, pp. 844, 909, 936, 1268. It is important to note that Grammatica Portugueza was printed in four separate parts by the missionaries at Tranquebar. The first one in 1725, the second one in 1726, the third one in 1727, and the fourth part in 1731 titled ' A Quarta Parte da Grammatica Portugueza (The Fourth Part of the Portuguese Grammar)' which presents um Vocabulario em Portuguez e Malabar, que contem os nomes, verbos e adverbios de ambas as lingoas, que são mais usadas no trato cotidiano (…). The three preceding parts shows an appreciation for the contrastive exercise vis-à-vis three languages: Portuguese, Tamil, and English. The fourth part includes other European languages besides Portuguese and English, namely: Danish, Dutch, and German.

23. Bibliotheca Nazionale Centrale di Roma, Vittorio Emmanule II (hereafter BNCR-VE), MSS Varia 53 (598), fls.85a -118b.

24. *Vocabulario Tamulico com a significacam Portugueza composto pello p. Antam de Proenca da Companhaia de Jesu, Missionario da Missam de Madurey, com todas as licencas necessaries da Santa Inquizicao e dos superiors, na impressa Tamulica do provincia do Malabar, por Ignacio Archamoni impressor della, Ambalacatta em 30 Julho 1679 annos*, see, the preface, para 2; Xavier S. Thaninayagam, *Antão de Proença's Tamil-Portuguese Dictionary, AD 1679*, Kuala Lumpur, 1966.

25. João de Sequeira was the son of Francisco de Negrereiros and Maria de Sequeira, born in Lisbon, baptized in the parish of S. Nicolão. He joined the Order of the Augustinian at the age of sixteen on 31 May 1673, professed on 3 June 1674, became Vice-Rector of Collegio do Populo de Goa in 1681, made Procurator-General of the congregation in 1684, went to Lisbon and returned to Goa in 1692, came to Mylapore and stayed in many places on the Tamil coast, went to Bengal in 1696 and returned to Goa, and became the novice-master on 21 September 1701, remained in the Convento da Graca in Goa in 1717 and died on 14 January 1724. António da Silva Rego, Documentação para a Historia das Missões, vol. XI, p. 550.

26. Bibliotheque Royale Albertine, Brussles, Belgium, MS no. 21868.

27. SCL, Goa, MS 37. *Vocabulario thamulico lusitano para uso dos missionarios da Companhia de IESV. Composto, e augmentado pello P. Domingos Madeyra da mesma Companhia, missionario na Missão de Madurey, Anno 1750*; See also, MS 39b, Pe. Domingos Madeyra, *Vocabulario thamulico lusitano para uso dos missionarios da Companhia de IESV. Composto, e augmentado pello P. ... da mesma Companhia, missionario na Missão de Madurey, Anno 1750*.

28. Ibid., MS 37, fl.3r. The manuscript is badly damaged and so transcription has been very difficult. The text in Portuguese runs as follows: Não he meo intento Religioso, e benevolo Leitor supplicar ao teu genio, que deixes de censurar neste Vocabulário, o que merecer censura, e repugnar ao teu prudente arbitrio; poes para ser o inferior, e mão reprehensivel, sempre nos bons talentos sobrou

justificada razão. So pretendo inculcar os motivos, que promoverão clara luz desta assaz limitada obra: são 1° ter sido rogado por varios P.P. fizesse este vocabulario pondo nelle muitas couzas, em outros faltão. 2°, por considerar, pa. satisfazer ao proveito como de tantos principiantes na lingoa thamulica, não ha nesta Prova. os vocabularies necessarios, e capazes; posto se tem escripto tantos por diversos A.A. em estillo mais elegante parto de tão subtiz engenhos como o foi proximamente o R. P. Constancio Joseph Beschi, o qual, sem exageração, não tem igual, assim no thamul vulgar, q chamão ...como no elegante, chamado...pa. a Poezia; e com tanta superioridade, excedeo aos mesmos Mestres naturaes, com razão se admirão ver em ...estrãgeiro tal sciencia. Pello ... atreverme eu a fazer esta obra, confesso ser acção digna de grande nota: motivo, por ...faço só huma couza corrente: na qual como a artificioza abelha,...das mais fragrantes flores colhe, ... só lhe aproveita pa. fabricar o doce favo em comum benificio não deixarei, cõ incansavel trabalho, de me aproveitar, e colher muitas regras, e documentos do sobredito P.: pa. cõ elles ordir ...como saboroso favo, ... suavize, e mova vontades ao estudo, e applicação do ... se contem neste vocabulario; no qual me persuado me não dilato de modo, ... cauze tedio, nem sou tão breve, ... ommitta obscuras duvidas, ...não declare; accomodandome ao proverbio, ...diz "omnia longa solent cunetis fastidi ...serve: quae comprehensa brevi sunt, bene cuncta placent" meyo na verdade proporcionado para attrahir o animo de todos â lição de qualquer tratado.

29. Ibid. fl.3r-3v. The text runs as follows: Pertence a Arte, nada escrevo; poes a idade de 65 annos mo não permite, e a pouca saude me impede o escrever tudo: mas supondo ja o estudo da Arte Thamulica, so trato dos vocabulos pertinentes ao Dicionario, no qual vem verbos, aos quaes asigno preteritos, futuros affirmativos, negativos, imperativos, infinitivos, e alguns participios, ... he o mais necessario aos principiantes: e na verdade o asignar preteritos aos verbos, he couza mais trabalhoza nesta lingoa, e so pello discurso de tempo se vem em conhecimento; e eu confesso, ... em espasso de trinta annos, ... tive de exercicio, sendo Missionario na Missão de Madurey, ainda em alguns vacillo; mas valendome de alguns peritos, satisfaço a minha pormessa. [fl. 3v] Com a graça de Deos espero satisfazer aos daqui e todos, pa. ... elles, como outro David, tendo a ajuda das palavras administradas, como escolhidas pedras na funda de sua eloquência, satisfação tal pontaria no Gigante da gentilidade, ... o possão derribar e qual outro santo, venhão a dares por vencidos, dizendo Domine quid me vis facere entregandose nas mãos dos Ministros da vinha do Senhor; ... que por sua devoção sejão eleitos como vazo, pa. elles tambem ser ...capazes instromentos de fazer conhecer aos demais o necessario pa. sua salvação. No fim vay um additamento, no qual ponho a sustancia da Introdução,... o R.P. Beschi faz no seo; poes julgo ajudarâ muito pa. a compozição, ler, e também escrever: e juntamente saber a força de algumas letras destes; e em ...cazos se podem pór singelas, ou dobradas; ou mudandoas em outras, segundo a exigencia, fiquem com a mesma significação.

30. HB, vol. I, continuation 8, pp. 505–605.

31. Anton Wilhelm Boehme, *Propagation of the Gospel in the East: Being a further account of the progress made by some missionaries to Tranquebar upon the Coast of Coromandel, for the conversion of the Malabarians; of the methods by them taken for the effecting of this great work; of the obstructions they meet*

with in it, and of the proposals which they make, in order to promote it. Together with some observations relating to the Malabarian philosophy and divinity: and concerning their bramans, pantarms and poets, London, Part II, second edition, 1711, see the letter of Ziegenbalg dated 22 August 1708, pp. 1-13.

32. HB, vol. I, continuation 9, 1716, pp. 662-808; vol. II, continuation 15, 1720, pp. 11-72; vol. II, continuation 16, 1720, pp. 73-151, and vol. II, continuation 17, 1722, pp. 153-224.
33. AFSt, M2 B7:14b. Albertine Gaur, Bartholomäus Ziegenbalg's Verzeichnis der Malabarischen Bucher, pp. 67, 86.
34. In 1712 George Lewis received a dictionary of Portuguese and English in palm-leaf. They wanted to make a fair copy of the same manuscript with collaborators. See, Arno Lehmann, ed., *Alte Briefe aus Indien: Unveröffentlichte Briefe von Bartholomäus Ziegenbalg 1706–1719*, Berlin, 1957, p. 301.
35. BNP, Paris, A.2487/1, pp. 2-8. 184
36. Biblioteca da Ajuda (hereafter BA), Lisboã, MSS Miscellanea Ultramarina, Codice 51-VI-30, fls. 200-05v.
37. Gabriel Francis Xavier Raj, *L'apport francais a la lexiocographie Tamile*, Jawaharlal Nehru University, Unpublished Ph. D Dissertation, 1995.
38. Julien Vinson, 'Note Supplementaire sur les Oeuvres Tamoules du P. Beschi', *Revue de Linguistitue et de Philologie Comparée*, vol. XXXIII, 1900, pp. 198-200.
39. British Library (hereafter BL), London, MS Or.1308, *Dictionaire Tamoul-Francois pour expliquer le tamoul vulgaire avec une preface ou sont plusieurs regles qu'i l faut scavoir avant de faire usage du dictionnaire compose par le Reverend Pere Constant Joseph Beschi de la compagnie de Jesus missionaire de l'Inde l'an 1744 ad majorem dei gloriam, virginis matris maria decus et religionis Christiana catholica romana Incrementum Amen*, see the preface.
40. Rigsarkivet (hereafter RA), Kobenhavn, Asiatisk Kompagnis, MS 1450b, *Vulgaris Tamulicae linguae dictionarium Tamulico-Latium additis in prefationae aliquot regulis necessario praelegendi*; JRULM, Tamil Manuscripts no.7; BL, European Manuscripts (hereafter MS EUR) D 104; SCL, Goa, MS M 15; BL, Additional Manuscripts, no. 26110, *Vulgaris Tamulicae Linguae Dictionarium Tamulico-Latinum*. This copy of Tamil-Latin dictionary was made by Paschoal Manoel dos Reys and he completed the work on 26 February 1750 and mentioned also in Tamil year date month and year. The day was Thursday as mentioned in the end. Pakkiam Pilley, ed., *Vulgaris Tamulicae Linguae Dictionarium TamulicoLatinum: Additis in Praefatione aliquot Regulis Necessario Praelegendis*, Typis South India Times Dictis, Trichnopoly, 1882.
41. SCL, Goa, MS 12, Pe. Constantio Josepho Beschio, *Dictionarium Tamulico-Latinum, 1744*, Part one of the dictionary, Tamil/Latin.
42. RA, Kobenhavn, Asiatisk Kompagnis, 1450b; BL, London, Additional Manuscripts, no. 26110. The Tamil text at the end of the London manuscript records: Mano vaakkku yetaatha nanmaiyai swabamaay kondirukkira paraapara vasthuvinudaya kirubai kadaashathinaalum yeppoththum parisuththa kanniyasthriyumaay namakku aandavalumaay irukkira archasishta maria nayagiudaya pirarthaniinaalayeum kaarana sargurukkaludaya paripurana

aasirvaaththainaallum yennudaya aasiriyargaludaya anugrahathinnaalyeum 1750 varusham maasi maatham 26, guruvaaramagiya viyaazhla kilami izhlamathiyaanathil intha agaraathiyai yezhuthi mudiththeyn, Paschoal Manoel dos Reys.

43. Madurai Province Jesuit Archives (hereafter MPJA), Shenbaganur, MS Tamil no. 4; Institut Français de Pondichéry Library (hereafter IFP), MSS DICT-TA, 13.1. The old inventory number is written at the left hand corner and it is 9767. This copy was made in the year 1806 from the version of 1744. In the first page it has been stated '*Dictionaire Tamoul a Francois pour expliquee le tamoul vulgaire avec une preface ou souz plusieurs regles qui l'faut scavoir avant de faire usage du dictionaire compose par le R(everend) P(ere) Constat Joseph Beschi de la compagnie de Jesus missionaire de l'inde l'an (1744)- 1806 ad majorem dei gloriam, virginis matris maria decus et religionis Christiana catholica romana Incrementum Amen*. There is a note in the beginning. N.B. Adcertas, quaeso, beuigue lector, ad duo subsequencia signa et. Stella praefixa vocabulis Tamulice scriptis illa e grandonica lingua faisse usurpa la denotal. Sagitta praefixa vocabuli cujuslibet Significationibus indicat illas in sublimiori lingua generatim vel poesi potius quam in usu quotidiano adhiberi. Hinc: 1. Omnes significations quae inveniuntur post praedictum signum, elegantiroes sunt et ordinarie mimine in usu a vulgo receptae.2. Significationes vero quae vel ante sagittam, vel sol & sine sagitta ponuntur, communis ut intellectae in lingua vulgari adhibentur, plermque saltem per Madurensis missionis ambitum. His ergo utere, si vis generaliter a populo intelligi, illas autem serva ad librorum vel academicorum intellctionem.

44. MPJA, Shenbaganur, MSS Tamil, no. 3, 5 and 9, *Vulgaris Tamulicae- Linguae Dictionarium, pars secunda in qua Lusitanis vocibus Latinae explicatis vulgaris idiomatis Tamulica verba adjunguntur*; See also, Old Manuscripts no. 225/5 fls.1-103; BL, London, Additional Manuscripts, no. 26110; JRULM, Tamil Manucripts no.11.

45. SCL, Goa, MS 12, Pe. Constantio Josepho Beschio, *Dictionarium Tamulico-Latinum, 1744*, Part two of the Dictionary, Latin/ Tamil; S. Jeyaseela Stephen, 'Portuguese Contribution to Tamil Lexicography', Indrani Mukherjee and Sovon Sanyal, eds., *Reading "Culture" in Spanish & Luso-Brazilian Studies*, New Delhi, 2004, pp. 354-77, see, pp. 372-4.

46. MPJA, Shenbaganur, MSS Tam, no. 5.

47. IFP, MSS DICT-TA 13.1, para 5.

48. Ibid., para 6.

49. Biblioteca de Sociedade de Geografia de Lisboa (hereafter BSGL), MSS Reservados, 146-A-17, *Vocabulario Portuguese, Tamul e Chingala, copeado por hum certo Missionario de Ceylão, Pe. de Congreagação do Oratorio de S. Philippe Neri et Goa o qual foi composto por Padre Jacome Gonsalves da mesma Congreagação e superior que foi et mesma missão et Ceylão*.

50. Arquivo Historico Ultramarino (hereafter AHU), Lisboã, MS Res. Man 1.

51. BSGL, MSS Reservados, 146(A) 16, *Vocabulario ordenado para os que se applicaco ao Tamil, e Chingala comecado pello Portuguez, proseguido pello Tamil, e accabado pello Chingala pello Padre Euzebio do Rosario da Congreagação do Oratorio de Goa anno de 1772*.

52. AFSt, M2 B6: 48; M2 B6:49 fls.1-6.
53. SCL, Goa, MS 32, Pe. Balthazar Esteves da Cruz, *Vocabulario Lusitano-Tamulico, 1738*, See also, *Catalgo dos Livros, Opusculos e Manuscriptos Pertences a Bibliotheca Nacional de Nova Goa*, Nova Goa, 1907, vol.1, p. 264.
54. Francisco do Carmo lived at St. Francis friary in Goa. His report of 20 November 1809 to the Viceroy of Goa described the activities of the Franciscans in Mylapore, Nagapattinam and Velangkanni. Historical Archives of Goa (hereafter HAG), Panaji, MSS Monções do Reino, livro 189, fl. 273; livro 191D, fl. 1161; livro 192B, fl. 539; livro 193A, fl. 1343, livro 195B, fl. 661 and livro 195c, fl. 310.
55. BSGL, MSS Reservados, 146 (A) 14, fls.1-144v. *Vocabulario Tamulico em Lusitano, Perture ao R.me Pe de Noronha Congragacao do Oratorio de cidade de Goa, Missionario Ilha de Ceylao, F.P. Pejalee 8 November 1838*. It has been recorded in folio one 'Archiv do Ministerio da Marinha e ultramar he 8 Fevreiro 1887 offerta Jose Joao Borges d Agore Ennes, Juiz de Ultramar'. The dictionary begins with the Tamil word Apa and gives Portuguese meaning baldada at folio1. Against the word varusham year we find 1627 to 1686 with Tamil names at fl.118. Word like Ramayanam is found in manuscript 92v. It may be mentioned that Antao de Proenca's Printed dictionary does not have this word. The year subhanu of 1703 and the year tharana had been mentioned as 1704 at fl.101. The year krodhi 1724 is mentioned at fl.89v; The year dundumi corresponds to 1682 is mentioned at fl.107. The names of Tamil years are mentioned along with the corresponding number with the present and future (1666-1716) at folio 56.
56. SCL, Goa, MS 33.
57. BSGL, MSS Reservados, 146 (A) 16.
58. SCL, MS 33, Anónimo, *Vocabulario lusitano-tamulico, séc. XVII*; SCL, MS 36, Anónimo, *Vocabulario lusitano-tamulico, séc. XVII*.
59. SCL, MS 38, Anónimo, *Vocabulario tamulico-lusitano, séc. XVIII*.
60. KB, Kobenhavn, MS Codice Tamul 31(UB), Phraseologia Tamulo-Arabica; KB, MS NY kgl.Saml. 154 4, Synonyma Ebraeo-Tamulica.
61. KB, MS Rostg.40 2
62. Albertine Gaur, Bartholomäus Ziegenbalg's Verzeichnis der Malabarischen Bucher, pp. 92-94; KB, NY Kgl.Saml.589C.
63. AFSt, HA 61:1.
64. SCL, MS 15, Padre Francisco de Noronha, *Significação de todas palavers no confessionario po mayor commoidade dos que começão a lingoa, 1890*.
65. BNP, MSS Indien, no. 224.
66. *Pingala Nikandu*, South India Saiva Siddhanta Pathippu Kazhagam, Tirunelveli, Second edition, 1978, noorpaa no. 2009.

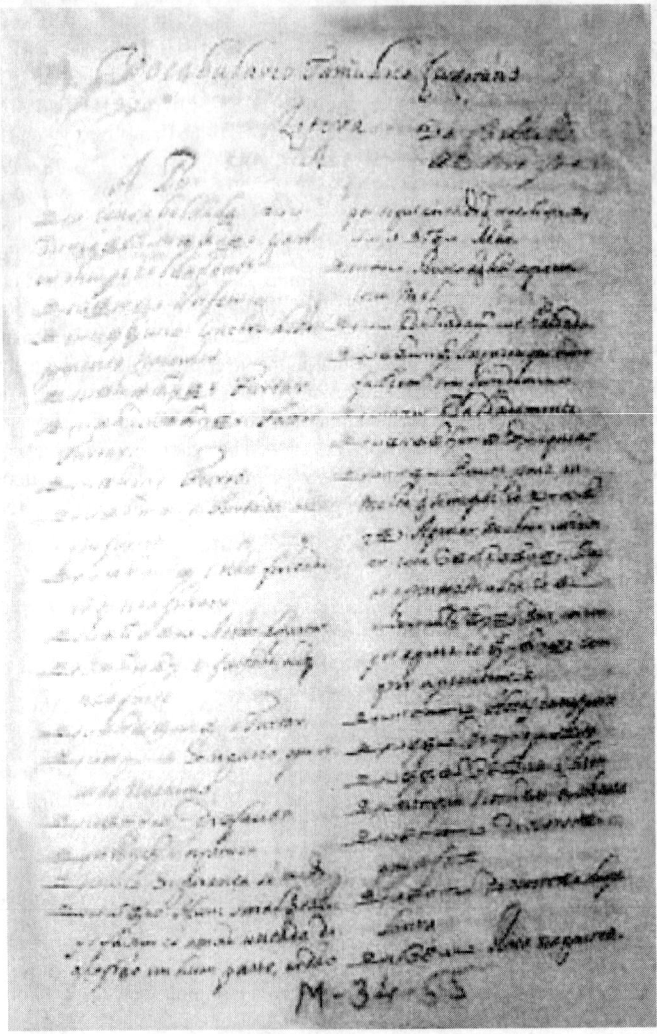

Vocabulario Tamulico

M-34-55

6.1. Tamil-Portuguese dictionary of Antão Proença, folio1 (Courtesy: State Central Library, Goa, MS. 34).

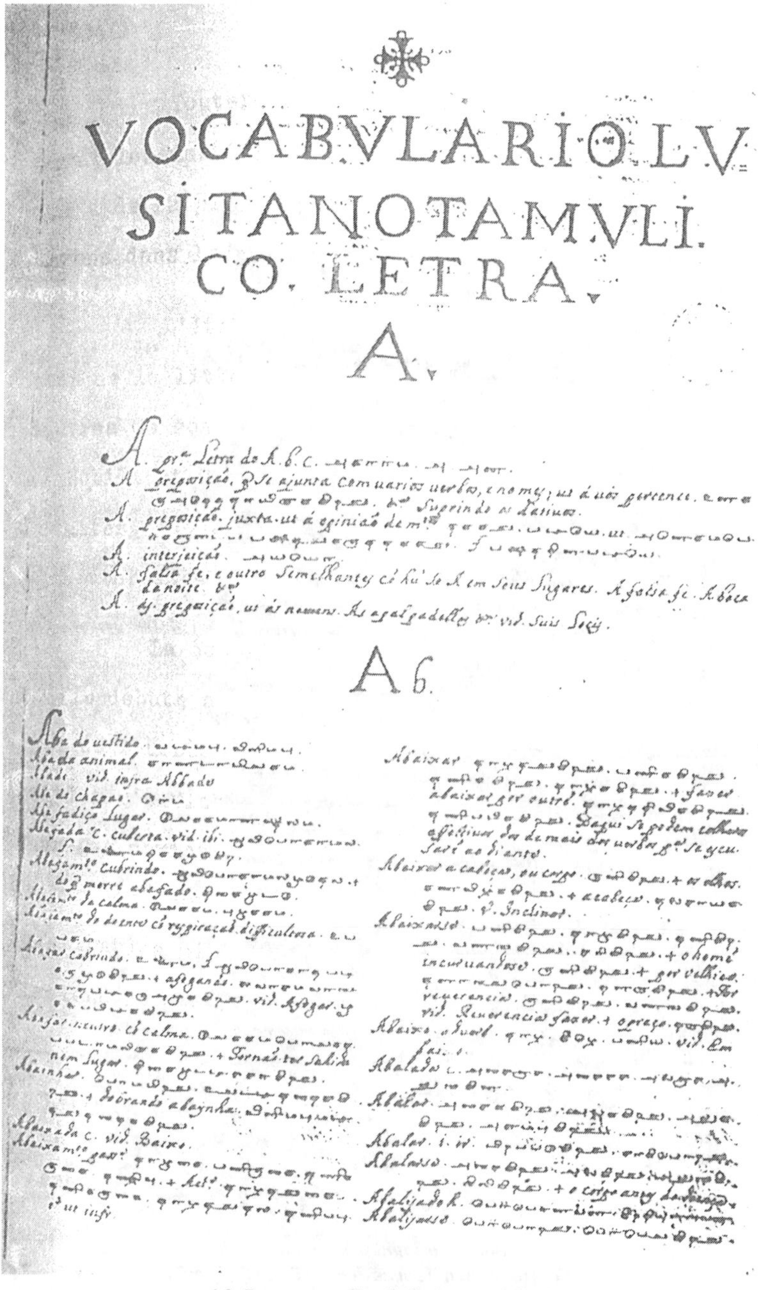

VOCABVLARIO LVSITANOTAMVLICO. LETRA.

A.

Ab.

6.2. Portuguese-Tamil dictionary, folio 1
(Courtesy: Bibliothèque Nationale, Paris, MSS Indien, no. 223).

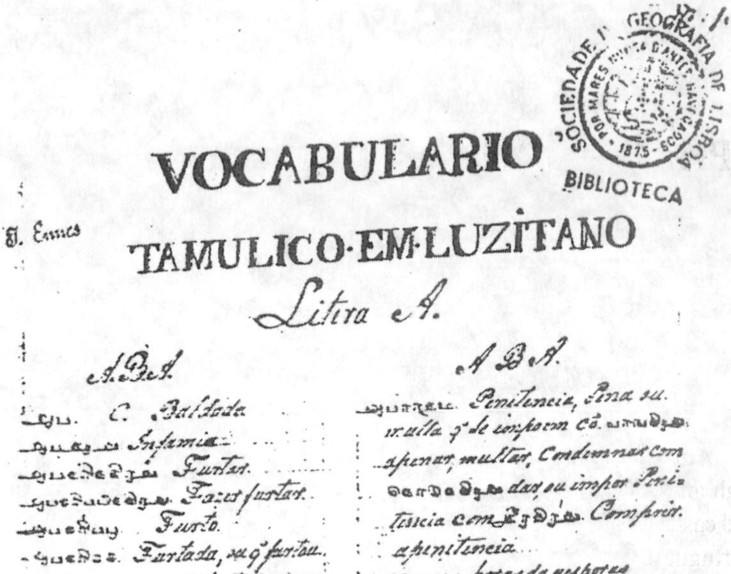
VOCABULARIO
TAMULICO·EM·LUZITANO
Litira A.

6.3. Tamil–Portuguese dictionary, folio 1
(Courtesy: Biblioteca de Sociedade de Geografia de Lisboa, Reservados, MS 146-A-14).

7

Printing Bi-Lingual Dictionaries and the Thesaurus of Tamil Language at Ambalakaadu and Madras for the Western Audience, 1679-1841

Antão Proença the Portuguese missionary had noted the existence of high and low Tamil, pure and corrupt Tamil, according to various classes and castes of the Tamil society during 1665 and 1679. He compiled a Tamil-Portuguese dictionary and testified the fact that Tamil Brahmins used Sanskrit words even in their ordinary conversations. A manuscript copy of his dictionary is available at the State Central Library in Goa.[1] It contains 16,546 main entries against which Tamil words and meanings have been given in Portuguese. It has 16208 non-lemmatized bilingual entries on 508 pages. In his preface to the dictionary he had provided guidelines for the Portuguese readers to read Tamil correctly by giving the transliterations of the Tamil words in Portuguese. Words given as subsidiary entries under one main entry have not been counted as separate entries. Hence the total nominal, verbal and adjectival besides composed forms and idioms in the dictionary if counted separately would increase the total of words more. He mentions that he had collected Tamil words from the dictionary of Ignacio Bruno (born at Bagnoli in the diocese of Nocera, Italy, joined the Society of Jesus in 1599 did missionary work in Jaffna Pattanam and died in 1659) but omitted words mostly used in the poetical works since they were not much in vogue in the spoken Tamil. Further Proença mentions that he had included many Sanskrit words in this dictionary since they were commonly used in the ordinary conversation of the Tamil Brahmins. He says that he used these Sanskrit words to help those who would read the books of Roberto Nobili and Manoel Martins, in which these Sanskrit words were found numerous.

Tamil-Portuguese Dictionary of Antão Proença Printed at Ambalakadu in 1679

Proença's dictionary begins with the Tamil word 'Aba' and gives meaning in Portuguese Cousa baldada, and also ends with the Tamil word 'shudir prayogam' giving the Portuguese meaning *Uso de cousasuim como feiticarias, embosses, onganos & c lit abuso*. His Tamil-Portuguese dictionary was the first one to be published by the Jesuit press at Ambalakaadu near Cochin on 30 July 1679, posthumously i.e. thirteen years after his death. The Tamil portions seem to have been made from engraved wooden blocks while the Portuguese portion has been made by using the moveable type. Fr Xavier Thaninayagam mentioned the printer was an Indian, Ignacio Aichamoni probably Accamani a Tamil name.[2] This opinion is called into question now for we find his name is mentioned as Patre Ignatio Archamone and also as Ignacio Arcamone in records who was a Jesuit. He was the manager of the college press of Rachol in Goa in 1663, and in 1670 his name figures in various publications.[3] However, the press at Ambalakaadu is called the Tamil press of the province of Malabar.

This printed dictionary original is available and preserved at the Vatican Library.[4] The order of the words runs: Aba, (cousa baldada), appoluthu pogirathu (gastar o tempo baldadamente); apathooru (infamia); apaththiyam (quebra do regimento); apagarikirathu (furtar); apagarikkapikkirathu (fazer furtar); apagarippu (furto); apagarichcha (cousa furtada); apagariyaathu (cousa nao furtada); and apgariththal (Accao de furtar). Frei Paolino de San Bartolomeo in 1792 had mentioned that the printed dictionary of Antão Proença was available in the library of the Society of Jesus in Rome. It may be mentioned that Gregory James wrongly attributes a copy of the manuscript of dictionary preserved in the library of the Sociedade de Geografia de Lisboa to Antão Proença.[5] In this dictionary the order of the words is different and it runs: Aba (C. Baldada); apathooru (infamia); apagarikirathu (furtar); apagarikkapikkirathu (fazer furtar); apagarippu (furto); and apagarikka (furtada).

There are three other different manuscript copies of Proença's Tamil-Portuguese dictionary in the National Library in Paris. One copy of the manuscript of Proenca's dictionary (under MSS Indien no. 221) has the dimensions of 22 x 34 cms, with a total of 197 folios and it was transcribed in the year 1750. This manuscript is written on both sides of the paper and each side of the page is numbered. It is gleaned from folio one that the dictionary was used by Fr John Ernest Hanxleden, known as Arnos Padre (1681-1732) in Malabar.[6] It is also recorded in folio two, that the manuscript dictionary was preserved in the beginning at the Royal Library of the King of France.[7] I have compared this manuscript dictionary along with the original printed version of Proença's dictionary in 1679 and surprisingly found that it is one and the same.[8] Hence it is concluded that Proença's dictionary was not reprinted, but missionaries continued to use the copies made in paper in India and it found way to France sent by the missionaries from Pondicherry.

There is also another manuscript copy of Proença's Tamil-Portuguese dictionary (under MSS Indien, no. 222) and it has the dimensions of 23 x 35.2 cms. It is written on both sides of the paper and each side of the page is not numbered. It is in double columns with a total of 111v folios. I have compared this manuscript dictionary along with the original printed version of Proença's dictionary in 1679 and surprisingly found that there is great variance.[9] The printed version was subjected to the censor board and so the omission of entries in the original are found.

The yet another manuscript copy of Tamil-Portuguese dictionary of Proença (under MSS Indien no. 224) is titled '*Dictionnaire Tamil-Portugais*' which is incomplete. It is recorded that this copy was made during the year 1848-9 by Mr. Raththinam. It has the dimension of 21.5 x 29.2 cms with a total of 103 folios and it is foliated in Tamil. A total of 5500 words are in Tamil in this dictionary. In folio number one, it is mentioned '*Copie le Tamil par Rattinam (1848) Le Portugais por invi (1849) sur un manuscrit appartenant a la famille Tiruvengadam, cet original est petit in 4 (vaqeir intrer) et se compose de 280 pagen don't la pra contrint de meme que la 3 de la 4, le commencement de l'ourage; le est comme la copie l' Indique, non termene, La garde de la relievre poste em n. 13 avec la signature Vazadarayar, Collatinni Par moi (1849) Edouard Ariel, Septembre, 1849*'. Both sides of the paper have been used for writing and page number is given only on one side. The dictionary is written in Tamil and Portuguese words are found only till page number eleven. Thereafter from page number eleven onwards, we find only Tamil words are written and the Portuguese words are incomplete. Tamil numerals are written up to folio 101 and later from folio 102 Arabic numerals are given. It is found that folios 11b to 102a do not have the Portuguese equivalents of Tamil entries. I have examined this manuscript and learnt that it is a copy of Proença's printed dictionary of 1679. The original of this fragment manuscript is even now preserved at the house of Ananda Ranga Pillai in Pondicherry. Mr Ananda Ranga Ravichandar, the descendant of the famous Tamil diarist, Ananda Ranga Pillai had been kind enough to permit me to have a look at this manuscript and use it for the purpose of research. In fact it is true that the transcription was made by Raththinam, one of the copyists employed by Edouard Ariel, who collected Tamil manuscripts in Pondicherry during the nineteenth century.[10]

Proença's introduction in the printed Tamil-Portuguese dictionary is very clear. We find here that as a compiler he was aware of his own limitations. The method followed by him in this dictionary compilation has been explained in the introduction. In this printed dictionary we find twelve reports and letters documenting the lengthy process of censoring the dictionary by the Catholic Church authorities in Ambalakaadu, Cochin and Goa during the four months period between 20 September 1678 and 6 December 1678. These no-objection certificates firstly relate to the teaching of the Catholic Church and secondly to the complete identity of Proença's original manuscript dictionary with the printed version. In this process four censors of Goa who are explicitly named as Tamil experts, give their opinion on the dictionary. The most important among these is João de Magalhaens, the procurator of the Jesuit province of Malabar and Goa who is described as reviser in Goa. On 21 February 1679 he writes that he has examined again the book called Tamil-Portuguese dictionary by Antão Proença which he used intensively for many years in studying Tamil language and practice. He states that he copied it with his own hand and used it for teaching and translation. The three more Jesuit Tamil experts are mentioned in the censor's section viz., Antonio Perreira at Ambalakaadu, Valeriano Caetano, also at Ambalakaadu and João Pessoa, professor of theology at the colleges in Goa and Cochin.

Proença listed lemmata in Tamil script but in Roman ABCD order, according to Portuguese phonological convention. Thus he begins with words in A, B/P, A, C/S in imitation of ABC instead of the Tamil order of independent vowels A, AA, E EE, followed by consonants. Proença has given his explanation and reasons for this order of entry, since he was preparing this vocabulary for the Portuguese missionaries and to facilitate several pronunciations for the same letter. Thus the dictionary work was aimed specifically for Portuguese speakers. On meeting a Tamil word, a Portuguese missionary could immediately look it up according to his perceived relation of sound and symbol, without having to learn the Tamil syllabo-graphemic order. This is thus a learner's dictionary constructed to the needs of the users. Colloquial idiom, a quick reference mechanism, head-words with subsidiary entries giving the multiple and incremented forms and a very brief individual Portuguese glosses for the Tamil entries.

Vocabulario Tamulico Lusitano

Letra

A

AB

Appa—Cousa baldada

Appozhuthu pogirathu—Galtar o tempo baldadamente

Abatooru—Infamia

Abathiyam—Quebra do reimento comer

Abgarikkiradhu—Furtar

Abgarippikkiradhu—Fazer furtar

Abagarippu—Furto

Abagarikka—Cousa furtada, ouque furtou

Abgariyaadha—Cousa na furtada, ou que nao furtoa

Abgarikkal—Accao de furtar

Abagariikira—Cousa furtada, ou que furta

Abgarikkugai—O furtar

Abagadam—Engano, q se Urde bo animo

Abagaaram—Disfavor

Abakeerthi—Infamia

Abaiyam—Seguraca se medo

Abayastham—Hum sinal, que os Reis faze co a mao untada co a safrao, em hum pano, e o dao por seguranca, do que o recebe ex

Abalam—Risco, a que hu sepoe. item, mal

Abam—Baldameto, ou cousa baldada

Abakkiyaathi—Injuria, que se dis salfamente se fundamento

Abayamaay—Baldadamente

Abakarikirigai—Exequias

Abaraatham—Penitencia, pena, ou multa, q se impoe; co- (vaangirathu) Apenar, multar, condenar-co – (kodukkirathu) Pagar a pena. A multa com. (kattaliidigirathu) Dar ou impor penitence.co. (thirkirathu) Coprir a penitencia

Abaaranam—Horas de uesporas

Abaththam—Despropofito

Abaththa visesham—Idem

Abavaatham—Descredito, defeza honra

Abasaanam—Descortesia q se fas

Abasaaram—Descortesia de palavra

Abaseyam—Rota na Guerra ou ser vebcido na Guerra, no jbgo xcz

Abaayam—Engano, ou mal que se fas; he o contrario, de (ubaayam)
Abaaya thanthiram, Abaayubaayam—Traca, ardilin malu
Abaanatha vaaram—Traseiro, modeste
Abaakkiyam—Mosina desgraca
Abaakiyan—H. mosino Abaakiyi—M. mosino
Abaavam—Nada, neggacao da realidade. gr
Abaiyam—Grito de quem pede justica.com. ritar, pedindo favor, ou justica com. – fazer justica &c
Abeshai—Desejo
Abeshichcha—C. que desejou, ou desejada
Abeshiyaaga —C.que nao deseja, ou nao he desejada
Abeshipikiradhu—Fazer desejar
Abeshippu—Desejo
Abeshikkira—C. que deseja, ou desejada
Abeshikkiradhu—Desejar
Abeshikkugai—O desejo
Abimandirichcha—C. benta.
Abimandiriyaatha—C. nao beta
Abimandiripikkirradhu—Fazer benzer
Abimanirippu—Renzimento
Abimandirikkira—C. que actualmente benzem, ou esta benzendo
Abimandirikkiradhu—Benzer rezando
Abimandirikugai—O bezer
Abhimaanam—A quillo deque hum se preza, ou aprezar se de aluma C. con (Indha sasthirathile yenakku abimaanamundu) – Prezome desta Siecia. Apud Xutres sinifica as partes secretas, modeste
Abhimanniyam—Pro.- na seunda significacao
Abinamarundha—Ansiao
Abippiraayam—Sentido, ou intento das palavras
Abirangak kaasu—Venezea no, moeda
Abipiadhagi—Augumento, incremento
Abisaari—M. adultera
Abishegam—Vneao
Abeesham—Desejo efficaz
Abirasa—C. rara, exquisite
Abhoorvam—Idem, fubst
Abhogam—False, q se impoem sem fundamento

Proença gives 65 words under 'AB' by giving the usage and the context. He also refers them as verb, pronoun etc., and other details and how it is to be used in grammar. The Tamil consonants in this printed dictionary have no dots or pulli. The 'r's do not have the oblique elongations. The long 'e' and 'o' are not distinct from the short 'e' and 'o'. The double loop (kombu) in Tamil has not been introduced. The cacuminal (r) has only one loop instead of the double loop as used in Tamil at present.

Proença mentions the purity and dignity of the Tamil words (*pureza e alteza de palavaras*) and he seems to have had a good understanding of both, high and the low, the pure and the corrupt forms in accordance with the usage. He had selected and used only such words of those words in the Tamil-Portuguese dictionary as would meet the requirements of the people in all walks of life to learn the Tamil language although it was mainly intended for the use of the Portuguese missionaries.[11]

A comparison between the manuscript original preserved in the National Library of Paris and State Central Library in Goa besides the printed version has very many interesting things to reveal. Here we find surprisingly the deletion included many passages of the preface as well as grammatical essays of Proença, parts which explain the transliteration of the Tamil head-word entries, and without which the alphabetical order of the dictionary cannot be properly understood had been omitted in printing. This had been the work of the censors who gave approval for printing after deleting. The word Ramayana had been explained in the original manuscript but the term has been deleted in printing. There are also some figures and drawings for words such as *Panchatsharam* in the original of Proença but these had been carefully avoided in printing for some reasons. It must be said that his innovative ideas of introduction and using diacritical signs to express vowel lengths in Tamil were ignored in typesetting possibly for technical reasons at the time of printing. The innovation leading to the modern Tamil writing system with diacritical marks and the historical importance and development by Proença had not therefore been made known to the researchers till date.

A careful study of the manuscript of Tamil-Portuguese dictionary preserved in Paris and Goa and its salient features needs to be introduced to the readers.[12] An examination of the printed version exhibits a lot of difference. There is a chapter on the alphabetical order used in the dictionary and a chapter on phonetics. Proença gives his reason for compiling the dictionary and describes the scope of the entries and their origins. He then shows the criteria for adding certain types of words and meanings to the dictionary and finally tells us what kinds of words have been left out. He leaves out many words spoken among the different groups of people in Tamil society especially those of the lower and rural castes and fishermen.

At the same time also the words that occur only in classical poetry are deliberately left out as they are useless for practical purposes and he says that there existed special dictionaries for that and he lists the works. Proença confesses that even after twelve years of studying Tamil he still had uncertainties about the meaning of some words especially as the native scholars themselves are not unanimous with regard to their meanings.

English-Tamil Dictionary of Fabricius Printed at Madras in 1776 and his Tamil-English Dictionary Printed at Madras in 1779 and its Revised Edition Printed in 1809

Johann Philipp Fabricius the Protestant missionary in Madras compiled a dictionary of English and Tamil and it was printed from the mission press at Vepery in Madras in 1776 in two volumes. The first volume of dictionary had 662 pages except the few preliminary pages and preface that had Roman numbers. The listing is done from the English alphabet 'A' till 'N'. The dictionary begins with the word 'A' and is explained as oru in Tamil. Then follows the words such as A man-oru manushan; A thing oru vasthu. Thus it is gleaned that meanings are given in Sanskritized Tamil. The listing in the second volume commences from the alphabet 'o' and it ends with 'z' and the last word is zodiac.[13]

Later we find that he continued the work and the dictionary had been improved upon and it was titled 'Dictionary of the English and Malabar languages' and it got printed at Vepery in Madras in 1786. The book has 272 pages.[14] It is interesting to note that dictionary was also produced by Johann Philipp Fabricius with the assistance of Johann Christian Breithaupt another Protestant missionary who also worked in Madras. The authors have mentioned that this dictionary work was meant for use in Europe and particularly England and also in India.

Fabricius also began to work on Tamil-English dictionary and he completed the work. It was printed and titled in Tamil '*Thamizhum Inglishumaayirukkira agaraathi*', along with the English title 'A Malabar and English dictionary, wherein the words and phrases of the Tamilian language, commonly called by the Europeans the Malabar language, are explained in English'. It was published from the Mission Press at Vepery in Madras in 1779. This is the first Tamil and English dictionary printed. It is significant to note that this was the dictionary wherein Fabricius had for the first time distinguished the Tamil words with Sanskrit roots with an asterisk mark. It was printed in two columns and no pulli (dot) was used over the consonants. Tamil alphabetical order was followed for words and the meanings are given in English.[15] This dictionary work was meant for the use of the people who wished to learn English in Tamil country.

A Tamil and English dictionary compiled by Rev. Fabricius and Breithaupt, the English missionaries at Madras published in 1779 had appeared as the second edition being revised and corrected by Rev. Poezold, Mr. William Simpson and the Tamil catechists. It was printed at Vepery in 1809.[16] It contains several names of plants in Tamil. Agathi the name of a tree whose leaves serve for pot-herbs had been mentioned and also we find its various types such as sevvagathai, another kind bearing red flower; sittagathi another kind having smaller leaves. Further, Aththi maram is mentioned as fig tree, while its other types such as vizhal aththi other kind of shrub; seemai aththi, a European fig tree had been mentioned. Alli thandu is mentioned as the name of a flower somewhat like a tulip and its other varieties such as sevvalli, the red kind and karualli, the blue kind had been mentioned. Adaalai maram is mentioned as name of a tree.[17] Thus plant names have been interestingly included in this dictionary.

The Thesaurus of Tamil Language titled Chaturagaraathi of Beschi of 1732 Printed at Madras in 1824 and Reprinted in 1835

Constanzo Giuseppe Beschi had remarked that Bourzés had produced a Tamil-French dictionary which evinced an amount of study, labour and zeal beyond all praise. However, as Bourzés had wished to mix together all the words of the ordinary and elevated dialects, and included them in his work without even omitting those terms used only by ignorant folk. He said that the dictionary of Bourzés had become so large that when one looks at it would be frightened and any young missionary would do away with learning Tamil.[18]

Beschi therefore at first decided to write a Tamil-Tamil dictionary dealing mainly with the peculiarities of the literary Tamil. This he titled in Latin *Thesaurus linguae Tamulicae ad pleniorum planioremque scriptorium Tamulensium intelligentiam collegit ac quatuor in partes digessit* and completed on 21 November 1732. This was composed for those who already knew Tamil.[19] It was called Chathuragarathi in Tamil by him being fourfold and divided according to words, synonyms, classes and rhymes. This was the first dictionary to have the words in Tamil alphabetical order. Beschi learnt directly the essential elements of the classical Tamil language from the traditional Tamil grammatical treatises. Beschi himself attested in his preface of Chathuragarathi that he had given each Tamil word with the meaning in Latin, Portuguese and French as he is very familiar with these four languages. It is also gleaned from his Latin preface of Chathuragarathi where he says that (*Vulgaris, Linguae Lexicon, ubi qualibet lectiones latine, Gallie, ac Lusitane explico*) he had given the explanations to each word in Latin, French and Portuguese in this common Tamil dictionary.

Following Agaraathi nikandu Beschi had used the term agaraathi for his work Chaturagaraathi. The reason is that Beschi had arranged alphabetically and introduced it newly in using the nikandus in Tamil that were available then for preparing this work. He says that it was a difficult task and he saw the old texts carefully, omitted the mistakes committed by the copyists of manuscript, corrected with the assistance of Tamil poets and scholars. He used the word thesaurus in Latin for this dictionary meaning treasury. In the first part of Chathuragaraathi under peyar agaraathi he gave 12187 words used and the meanings. In the next part under porul agaraathi he gave many words that gave single meaning and compiled the list. In this he gave a list of 2168 words.

All the meanings of a single word had been given by Beschi in his Chathuragaraathi. Those words which refer to God commonly such as aathi, karuthaa, karuthan, munaivan, moovulagunarnthon, pooranan, vimalan, aththan, anaathi, suththan, suyambu, sami, swami, jothi, arubi, yegan, Isan, Naayan, Naathan, Devan, paran, paraman, vinnavan and perumaan are mentioned by Beschi. The meanings of some words such as mannan-mannavan, panri-varaahan, unavu-unaa, uvamai-uvamam, suththam-susi, umbi-numbi and vizhla-vizhavu are really surprising to us. He noted many same words which gave the same meaning such as Aruththa, umi, uri, yerumai, kudirai, kanu, killal, kizhithal, kudal, kudukkai, kuyavar, kurudu, kuruththu, virudhu, viraghu, vellam and vellaadu. He mentioned the short and the long (kuril and nedil) differentiation in the pronunciation of letters and gave it nicely. He also made a study of kurukeezhilethugai and nadir keezhilethugai and separately assembled them under different headings, which is the beauty of Chathuragaraathi.

Beschi had mentioned several terms related to sickness and disease in his Chaturagaraathi such as aralai, akkaram, ananthar, unmaththam, kapam, kazhalai, kaandagam, sanni, sanangu, piththam and kuttam. Thus this Tamil dictionary has been prepared with the total of 21,870 words for the use of those who had already learnt Tamil with a view to serve as a guide for improvement and developing their skill in the language. It was printed at Madras in 1824.[20] It was reprinted by Rev. J. Smith at the Church Mission Press at Madras in 1835.

Printing the Dictionary of the Tamil and English Languages Compiled by Johann Peter Rottler at Madras, 1834-41

Johann Peter Rottler (1749-1836) the Lutheran missionary arrived at Tranquebar on 5 August 1776 and worked there till 1803. He was much interested in botany and made several trips for collecting plants. Later he settled at Madras and worked there until his death in 1836.[21] Rottler compiled a Tamil-English dictionary and it was published under the title 'A Dictionary of the Tamil and English Languages' in Madras in 1834. This language dictionary published is a kind of botanical dictionary containing the new scientific names of plants along with Tamil names in it. There are generic names, new specific epithets and new varieties of plants in Tamil country. Several specific epithets have later homonyms too. It is a dictionary of medically and economically important Tamil plants both beneficial and deleterious arranged in alphabetical order.

Rottler had special interest in plants beyond his task of recording what others had said. His knowledge of botany is demonstrably strong. Most of the entries have scientific names of plants. The plants named by botanists like Linnaeus, Vahl, Roxburgh, Willdenow and Retzius are found. Equally interesting is the fact that the word '*nobis*' is used by the group of scholars who had given plant names. Again only the genera had been mainly found mention in many cases of plants. Many botanical names as mentioned by Linnaeus had been given. While in some of the cases the genus and the species had been also given. Rottler had clearly mentioned the various varieties of the same plants under the different names. In order to make the English reader understand he mentioned the plant types and related details such as trees, shrubs, plant, grass, seed, fruit, etc. The same plant with different names used by Tamils under colloquial and literary forms had been given. In some cases various species had been mentioned. In some cases only Tamil names are mentioned and reported as a plant without aiming to provide the scientific name. Equivalent English names of the plants in Tamil had been given to facilitate the Tamil reader. The selection of the Tamil alphabet 'Ka' alone shows the following details of plants in English and the botanical names.[22]

1. Kak maram - same as manathakkaazhi a plant - Solanum
2. Kakari - a kind of cucumber - Cucumis muricatus
3. Kasakasa - a plant the poppy - Papaver
4. Kasangu - a palm - Phoenix
5. Kasamaathu - Umaththai, a plant - Datura
6. Kachi - the same as seenthil, a winding plant - Mensipermum

7. Kachuram - the same as kalrchikodi, a shrub - Guilandina
8. Dumbashdagum - Alpinia galanga major
9. Kandupirangi - another medicinal root
10. Kanja - an intoxicating plant - Cannabis
11. Kanjam - the same as thaamarai a plant - Nelumbium
12. Kanjankorai - a medicinal plant - Ocimum album
13. Kaatu kanjangkorai - a different species- Ocimum fastigatium, Nobis.
14. Sirukanjangkorai - another species - Ocimum ciliare, Nobis
15. Kadambum - a plant - Nauclea cadamba
16. Kadambal - the same as kumizh, a tree - Gmelina
17. Kadambu a tree - Eugenia racemosa also called neerkadambu as it grows generally near water.
18. Nilakadambu - a plant - Insticia acaulis.
19. Venkadambu - a variety of the kadambu bearing white flowers.
20. Kadalai - Bengal gram - Cicer arietium
21. Verkadalai - groundnut of the West Indies, Arachis hypogoa
22. Kadalinjigum - the same as kusai pul, a grass - Saccharum
23. Kadalakki - the same as pey musuttai, a plant
24. Kadalaadi - a plant - Achyranthes
25. Kadal viranji - a shrub
26. Kadarkodi - the same as thumbai a plant - Phlomis
27. Kadarpaalai - a plant - Convolvulus speciosus (samuthirasogi)
28. Kadavu - the name of a tree thankku – Gyrocarpus
29. Kadaarai - kadaar naarthai, a citron tree - Citrus
30. Kadikandu - the same as punaikaali, a plant - Dolichos
31. Kadichchai - the name of a shrub - Caesaria elliptica
32. Kadiyirtham - the same as mukkirattai, a plant - Boerhaavia
33. Kadi romam - the same as koraikilangu, the root of Cyperus
34. Kadu - a tree - Terminalia
35. Kadukkaay - the medicinal fruit of the Terminalia chebula
36. Kadugam - the same as kadurogini a plant - Helleborus
37. Kadugu - a plant - Sinapis chinensis - mustard
38. Siru kadugu - Sinapis racemosa - Nobis
39. Naay kadugu - a plant - Cleome viscose- dog mustard

40. Malai kadugu - hill mustard
41. Ven kadugu - white mustard
42. Kadugu - the same as kunri, a shrub - Abrus
43. Kaduppadaki - the same as Yerumuttai Pee Naari, a plant - Ballota
44. Kadukkai - the same as konrai maram, a tree - Cassia
45. Kadukkai - the same as maruthu, a tree - Terminalia
46. Kadudum - the same as marukkaarai, a shrub - Gardenia
47. Kadaicha - the same as netti, a shrub - Aeschynomene
48. Kattanchi - a tree - Celastrus emarginatus
49. Kattu kodi - a shrub - Smilax pseudochina
50. Perung kattu kodi - Smilax tomentosa
51. Kanappoondu - a medicinal plant - Exacum pedunculatum
52. Kanamuli - the same as Saalukki, a medicinal plant - Ocimum
53. Kanaviram - the same as alari, a shrub - Nerium
54. Ponnangkanni - Illecebrum sessile
55. Karisalangkanni - Eclipta erecta et prostrata
56. Kangaum Neermulli- a thorny plant - Barleria
57. Kanga sangam - a thorny shrub - Monetia
58. Kandangkathiri - a medicinal plant - Solanum jacuini
59. Kandal - a tree - Rhizophora candel
60. Kandal - the same as thaazhai, a tree - Pandanus
61. Kandal - the same as mulli, a plant - Solanum
62. Kandal - the same as neermulli - Barleria longifolia
63. Pushpa kandal - candel tree bearing fragrant flowers
64. Pey Kandal - a tree - Rhizophora mangle
65. Kandu mulam - the same as kandupongi, a medicinal plant
66. Kandugam - the same as manjiti, a plant
67. Kandu pangi - the same as siru thekku
68. Kanduthi - the same as kanjori, a plant - Tragia
69. Kanni kodi - a winding plant - Apocynum frutescens
70. Kannigam - the same as manathakkaazhi, a plant - Solanum
71. Kathambum - the same as kadambu, a tree - Eugenia
72. Kathambai pul - a kind of grass with twisted awns - Andropogon
73. Kathali - the same as - thettra maram a tree - Strychnos
74. Kathali - the same as vaazhai, a superior kind of plantains

75. Kathiram -the same as karungkaali, a tree - Diospyros
76. Kaththari - a plant, brinjal - Solanum melongena
77. Kotthu kaththari - a plant- Solanum stramonifolium
78. Kaaral kaththari - a plant - Solanum insanum
79. Kathigai - the same as karukkuvaazhi, a tree - Zizyphus
80. Kantha satgum - the same as thamarthai, a tree - Averrhoa
81. Kantha puthiyam - the same as naay vezhai, a plant - Cleome
82. Kantham - the same as karnai kilangu, a kind of root - Dracontium
83. Kantharum - the same as punal murungai, a plant - indigofera
84. Kanthi vaaruni - the same as peythumatti, a plant - Bryonia
85. Kanthugani - the same as thannri, a tree - Terminalia
86. Kantheru - the same as kodagasaalai, a plant - Justicia
87. Kabitham - the same karunjeeragam, a kind of seed - Nigella
88. Kapitham - the same as vizha maram, a tree - Feronia
89. Kappadi maram - a medicinal tree
90. Kamalam - the same as thaamarai - Nelumbium speciosum
91. Kamalai - the same as kudasapaalai, a plant - Asclepias
92. Kamugu - the betel nut tree, a palm - Areca Catechu
93. Kunthar makugu - a palm - Caryota urens
94. Kampali kondaan - a shrub - Morus indica
95. Kampu - a kind of grain - Holcus spicatus
96. Kammai - the same as siru keerai, a plant - Amaranthus
97. Kayirvam - the same as vellambal, a water plant - the white Nymphoea
98. Karunai - a plant - Dracontium
99. Karrakari karunai - another species - Tacca Pinnatifida
100. Kaatukkarunai - another species used in medicine - Dracontium polyphyllum
101. Karathai - a plant - Ocimum basilicum
102. Kuthu Karanthai - a plant - Erigeron Obliquum
103. Kottai karanthai - a plant - Sphoeranthus Indicus
104. Sirukottai karanthai - a plant - Ethulia divaricata
105. Siva karanthai - a plant - Sphoeranthus zeylanicus - Nobis
106. Sunai karanthai - a kind of Ocimum - Ocimum pilosum - Nobis
107. Surya karanthai - a plant

108. Narung karanthai - the same as siva karathai so called form its strong smell
109. Vishnu karanthai - the same as kottai karanthai
110. Karisalangkanni - a plant and medicine for the eyes, Eclipta erecta
111. Karippan - a plant
112. Kareerum - the same as agathi a tree - Coronilla
113. Karugkaali - a tree – ebony - Diospyros Ebenum
114. Karunguvalai - the same as Neelorpavan, a water plant - Pontederia hastata
115. Karungkodi - a twining shrub - Smilax pseudochina
116. Karuvilam - the same as kakkanam, a winding and leguminous plant - Clitoria
117. Karuvilam - a tree - Aegle marmelos
118. Karaa mani - a leguminous plant - Dolichos catiang
119. Karudathondai - the same as kakanaang chedi, a plant - Clitoria
120. Karudathondai - the same as kovai kodi, a plant - Bryonia 121. Karpura maram - the camphor tree - an oak
122. Kalappai kilangu - same as karthigai kilangu, a plant - Gloriosa
123. Kaligai - the same as Naagamalli, a shrub - Justicia
124. Kalingam - the same as vetpaalai, a plant - Nerium
125. Kalisum - the same as vanni maram, a tree - Prosopis
126. Kalithurumam - the same as thannri maram, a tree - Terminalia
127. Kallukalai kaathaan - the same as ponnagkanni, a plant - Illecebrum
128. Kalluruni - the same as kuruvichi, a shrub - Ehretia
129. Kallarum - the same as sengaluneer, a water plant - Nymphoa
130. Kallarum - the same as Neerkumizhi, a water plant – Damasorium
131. Kalangu - the same as kalr kodi, a shrub - Guilandia
132. Kalal - a shrub - Guilandia bonduccella
133. Kalanthoori - the same as thanrikkay, the fruit of a tree, Terminalia
134. Kazhaa, a shrub - Carissa
135. Siru kazhaa - Carissa spinarum
136. Perung kazhaa - Carissa carandus
137. Sothai kazhaa - a shrub - Flacourtia sepiar
138. Kazhaavagam - the same as siru keerai, a kind of greens - Amaranthus

139. Kalli - the name of a shrub, milk hedge, Euphorbia tirucalli
140. Attang kalli - the same as thirukalli - Euphorbia tortilis - Nobis
141. Ilai kalli - a twining plant - Cynanchum viminale
142. Sadura kalli - another species - Euphorbia antiquorum
143. Palagai kalli - a plant - cactus - Ficus indica
144. Perung kalli - the same as Izhathalari plumieria
145. Maan sevi kalli - a shrub - Cacalia Kleini
146. Kanag kaay - a tree - Uvaria Kananga
147. Kanni kaaram - the same as kongu, a kind of trees- Bombax
148. Kannirum - the same as isangu, a thorny shrub- Monetia
149. Kannurugam - the same as sirukeerai, a plant - Amaranthus

Thus Rottler in his dictionary of the first volume, part one, had brought out the previously used scientific names in Europe to the attention of the botanical community to prevent the creation of additional homonyms. The Tamil-English Dictionary of Rottler volume I, Part II was printed in 1836.[23] Then the Tamil-English Dictionary of Rottler volume I, Part III was revised by W. Taylor and T. Vencatachala Moodelly and it was printed in 1839.[24] Finally, the Tamil-English Dictionary of Rottler volume I, Part IV was also revised by W. Taylor and T. Vencatachala Moodelly and it was printed in 1841.[25]

It may be mentioned that all the four volumes (vol. I, 302 pages, vol. II, 421 pages, vol. III, 457 pages and vol. IV, 254 pages) had together 1411+xxiv pages amounting to a total of 1435 pages printed between 1834 and 1841. Thus Rottler's dictionary was successfully printed by the Protestant missionaries in Madras.

To conclude it may be said that the substance of the dictionary printed by Proenca in 1679 is that it had Tamil-Portuguese homonyms while that of Fabricius it had Tamil-English homonyms. The substance of the Chathuragaraathi is that it had Tamil-Tamil homonyms, synonyms, group names and rhyming words. It has been also noted that the dictionary of Antão Proença and Constanzo Giuseppe Beschi the Catholic missionaries were printed after subjected to by much scrutiny and approval by the Church authorities in Goa and the Bishop of Mylapore. They have deleted many things which the authors submitted in the original. Protestant missionaries like Johann Philip Fabricius and Johann Peter Rottler did not face such problems because they were free from getting the approval for printing. Many dictionaries prepared by the Catholic missionaries therefore had remained in manuscript form but got widely circulated.

Notes

1. State Central Library (hereafter SCL), Goa, MS 34, Antão de Proença, *Vocabulario Tamulico Lusitano precedido de uma Arte em Tamul e seguido da termos poeticos*, dated 1670. There is another copy of this dictionary that exists but undated. SCL, MS 38. The dictionary is titled *Vocabulario Tamulico-Lusitano* with 223 folios in two columns.
2. *Vocabulario Tamulico com a significaçam Portugueza composto pello p. Antam de Proença da Companhia de Jesu, Missionario da Missam de Madurey, com todas as licencas necessaries da Santa Inquizicao e dos superiors, na impressa Tamulica do provincia do Malabar, por Ignacio Archamoni impressor della*, Ambalacatta em 30 Julho 1679 annos, see, the preface, paragraph 2; Xavier S. Thaninayagam, *Antão de Proença's Tamil-Portuguese Dictionary, AD 1679*, Kuala Lumpur, 1966; Xavier S. Thaninayagam, 'Antão de Proença's Tamil-Portuguese Dictionary, 1679, An Introduction', *Tamil Culture*, vol. XI, no.2, April-June 1964, pp, 11727, see p.121.
3. Biblioteca Casanatense, Roma, *Purgatorii Commentarium Concannice Compositum*, Goa, 1663; Bodleian Library, Oxford, *Conciones per Annum Concannice Compositae a Patre Ignatio Archamone e Societatis Iesu*, Goa, 1670.
4. Biblioteca Apostolica Vaticana (hereafter BAV), Città del Vaticano, Borgiano Indiano 12; Edgar C. Knowlton, 'Antão de Proença's Vocabulario Tamulico Lusitano: Indo-Portuguese Elements', *Tamil Culture*, vol. XI, 1964, pp.135-64; E.C. Knowlton, 'Portuguese-Tamil Linguistic Contacts', in *Proceedings of the First International Seminar-Conference of Tamil Studies*, 1966, pp. 607-17.
5. Gregory James, *Col- Porul: A History of Tamil Dictionaries*, CREA, Chennai, 2000, p. 598. The manuscript preserved at the Geographical Society of Lisbon under 146-A-14 (Res) is not the work of Antão de Proença. But this manuscript dictionary is the work of Fr. Paulo Francisco de Noronha as stated in it. It was a copy made at Pejalle (Pesalai at Thalaimannar) on 8 November 1838 and it is mentioned at the end.
6. Bibliothèque Nationale, Paris (hereafter BNP), MSS Indien no. 221. In folio one it is written, 'auctor forte R.P. Hanxleden vide Exam. Historcrit calendre Bibl. Cong. do Propaganda Fide, ouet, P. Paulion (1792) p.55 no. X'.
7. Ibid., In folio two it is written 'MSS Inde de La Bibliothèque du Roi no. 204'.
8. Ibid., fls.1-197. The dictionary first page starts with the word Apa and ends with the word Achu, while the rest of the folios continue to run. Against the word varusham in this manuscript at folio 161 the years are not mentioned. So also in the printed dictionary against the word varusham at page 209 no years are furnished.
9. BNP, MSS Indien, no. 222, fls.1-11v. The dictionary folio one starts with the word Apa and ends with the word Abimaanam. The folio two of the dictionary starts with the word abimanyam and ends with the word athitti, while the rest of the folios continue to run. Against the word varusham in this manuscript at folio 91 the years are not mentioned. So also in the printed dictionary against the word varusham at page 209 no years are furnished.
10. S. Jeyaseela Stephen, *The Diary of Rangappa Thiruvengadam Pillai, 1761-1768, Translated from Original Tamil*, Pondicherry, 2001, p. xv.

11. BAV, MSS Borgiano Indiano, no. 12, fl.10.
12. Thomas Malten, 'Antam de Proença's Tamil-Portugucse Dictionary, Printed in Índia in 1679', in Charles J. Borges and Helmut Feldmann, eds., *Goa and Portugal: Their Cultural Links*, New Delhi, 1997, pp. 271-79. See also, Gregory James, 'Aspects of the Structure of Entries in the Earliest Missionary Dictionary of Tamil', in *Missionary Linguistics IV/Linguistica Misionera IV, Lexicography: Selected Papers from the Fifth International Conference on Missionary Linguistics, Merdia, Yuctan, 14-17 March 2007,* eds., Otto Zwartjes, Ramon Arzápalo Marín and Thomas C. Smith-Stark, Amsterdam, 2009, pp. 273–302.
13. Archives of the Gurukul Lutheran Theological College and Research Institute (hereafter AGLTC), Chennai, Microfilm, Accession Number 578, Roll no. 51, Johann Philipp Fabricius, *A Dictionary of English and Malabar*, II vols., Mission Press, Vepery, Madras, 1776; Gerald Duverdier, 'L'Imprimerie tamoule de vepery et es deux dictionnaire de Fabricius', Gutenber-Jahrbuch, 1978, pp. 189-97.
14. School of Oriental and African Studies Library, London, EB.78.177, EB.78.29; BL, London, 621.1.29; KB, Kobenhavn, 72-1113; Royal Asiatic Society Library, London, 175B; Gottingen University Library, Ling.II.7345, Johann Philipp Fabricius and Johann Christian Breithaupt, *A Dictionary of the English and Malabar Languages, Printed and Completely done at Vepery, near Madras,* 1786.
15. Bodleian Library, Oxford, Drave.d.4; Library of the School of Oriental and African Studies, London, EB.77.16; BL, Library, London, 622.k.9; 621.1.29; Gottingen University Library, Ling.II.7345.1, Johann Philipp Fabricius, *A Malabar and English Dictionary, Wherein the Words and Phrases of the Tamilian Language, Commonly called by the Europeans the Malabar Language, are Explained in English by the English Missionaries of Madras*, Wepery, near Madras, 1779. See, P.R. Subramanian, 'The Metamorphosis of Tamil-English Lexicography: Dictionaries of Foreign Missionaries in the Period 1750-1900', in C.T. Indra and R. Rajagopalan, *Culture, Language & Modernity: English-Tamil in Colonial India, 1750-1900*, Abingdon, 2018, pp. 30-47.
16. Fabricius and Breithaupt, *A Malabar and English Dictionary Composed by the Rev. Messr. Fabricius and Breithaupt, English Missionaries at Madras and Published in the Year 1779*; The Second Edition Revised and Corrected by the Rev. Mr. Poezold, Mr. William Simpson and the Malabar Catechists, Printed at Vepery in the Year 1809, Madras, 1809.
17. Ibid., pp. 1, 3, 5, 7.
18. BNP, MSS Indien, no. 215, see the preface of the Tamil-French Dictionary of 1744.
19. SCL, Goa, MS 11, *Chathuragaraathi: Quadruple dictionnaire Tamoul de Beschi, Contenant le Dictionaire des Mots, Celui des Synonymes, Celui des Classifications et Celui des Rimes*, 1741. This Tamil dictionary with words of Senthamizh, a fourfold dictionary of High Tamil, giving (1) meanings (2) synonyms (3) classes and (4) rhymes and it is dated 1741. It suggests strongly that a copy from the original of Beschi in 1732 was made. BL, London, no. 14172 f.13.
20. *Chaturagaraathi: Thesaurus Linguae Tamulicae ad Pleniorum Planioremque Scriptorium Tamulensium Intelligentium Collegit ac Quatuor in Partes Digessit*,

eds., V. Thandvaraya Mudaliyar and Ramchandra Kavirayar, College Press, Madras, 1824.

21. S. Jeyaseela Stephen, *A Meeting of the Minds: European and Tamil Encounters in Modern Sciences, 1507-1857*, Delhi, 2016, pp.134-139.
22. *A Dictionary of the Tamil and English Languages by J. P. Rottler, Missionary of the Society for the Propagation of the Gospel in Foreign Parts*, vol. I, Part I, Madras, Printed at the Vepery Mission Press, 1834, pp.1- 65.
23. *A Dictionary of the Tamil and English by the Late Reverend J. P. Rottler Missionary of the Society for the Propagation of the Gospel in Foreign Parts*, Vol. I, Part II, Madras, Printed at the Vepery Mission Press, 1836.
24. *A Dictionary of the Tamil and English by the Late Reverend J. P. Rottler Missionary of the Society for the Propagation of the Gospel in Foreign Parts, Part III, Revised by W. Taylor and T. Vencatachala Moodelly*, Printed by K.W. Thrope, at the Vepery Mission Press of the Society for Promoting Christian Knowledge, 1839.
25. *A Dictionary of the Tamil and English by the Late Reverend J. P. Rottler, Part IV, Revised by W. Taylor and T. Vencatachala Moodelly*, Madras, Printed at the Vepery Mission Press, 1841.

VOCABVLARIO
TAMVLICO.
COM A SIGNIFICAÇAM
PORTVGVEZA.
COMPOSTO PELLO P. ANTAM DE PROENÇA,
DA COMPANHIA DE
IESV
MISSIONARIO DA MISSAM DE MADVREY.

Com todas as licenças necessarias da santa Inquiziçaõ.
E dos Superiores.

Na impressa Tamulica da Prouincia do Malabar, por Ignacio Archamoni impressor della, Ambalacatta em 30. de Iulho, 1679 annos.

7.1. The title page of Tamil-Portuguese dictionary of Antão Proença printed at the Jesuit Press in Ambalakadu on 30 July 1679 (Courtesy: Biblioteca Apostolica Vaticana, Città del Vaticano).

7.2. The last page of Tamil-Portuguese dictionary of Antão Proença printed at the Jesuit Press in Ambalakadu in 1679 (Courtesy: Biblioteca Apostolica Vaticana, Città del Vaticano).

A DICTIONARY

ENGLISH AND MALABAR.

A — ABI

ABI — ABO

7.3. English-Tamil Dictionary by Johann Philipp Fabricius printed at Vepery in Madras in 1786, page 1 (Courtesy: Universitats Bibliothek Gottingen).

8

Louis Savinien Dupuis in Pondicherry and Printing the Tri-Lingual and Bi-Lingual Dictionaries for the Tamil Audience 1846-75

The first French-Tamil and Tamil-French dictionary was published in Paris in 1831 by André Blin, the Lieutenant of the Fourth Infantry Regiment of Pondicherry and Member of the Asiatic Society of Paris. It has 281 pages.[1] A close look at the lithographed book shows full of errors even in French language. The MEP missionaries in Pondicherry desired to introduce printed books in the seminary for teaching and use. Their primary need had been considered very important for having decided to the train the native seminarians for priesthood. In the first place, the seminary requirement and demand was the composition of all the classics and most of all the spelling books and an abridged Tamil grammar. Priority was also assigned to have a Latin-Tamil dictionary printed. In the first synod of Pondicherry held in 1844 it was decided by Clement Bonnand the Bishop of Pondicherry and the missionaries that immediately after composition of the works needed for the seminary they wished to proceed to the printing of the same.[2] The formation of the native clergy had formed an important matter of subject in the synod.[3] It was also discussed at that time as to how the seminary and the school would function and what books to be printed for use. The minor seminary teaching and the medium and method of instruction was well discussed in the same synod.[4] The educational plan was thus chalked out and it emphasized the teaching of Latin language and included various other subjects such as rhetoric following the method which experience had proved to be the best efficacious for ensuring the progress of the students. The Tamil language for its natural and very proper was emphasized so that the young students could learn their mother-tongue correctly. This was felt highly necessary to the seminarians for they could have greater respect in the eyes of the non-Christians, so that people would have no scorn for a Catholic priest who knew his own language perfectly. The knowledge of French had been also felt necessary because it would be a fecund source of religious and classical instruction and would help in making good rapport between the European priests and the native clergy.

The Attempts to Compile Latin-French-Tamil-English Dictionary by Louis Savinien Dupuis in Pondicherry, 1841-2

The making of dictionaries for the use of native seminarians in the Pondicherry mission gained priority by the MEP missionaries. Louis Savinien Dupuis the missionary launched the work and he mentioned in his letter of 18 August 1841 that he seriously envisaged the compilation of Latin-French-Tamil-English dictionary. In his letter addressed to His Grace Bonnand he wrote that the Bishop knew that not a single complete dictionary printed had existed for use in the city of Pondicherry. The seminary was in a dreadful state of stagnation because of the lack of a printed dictionary. Dupuis added that in India there are missionaries who spoke French, English, and Latin, and the seminarians too needed to be taught Latin. Hence he had planned to compose a Latin-French-Tamil-English dictionary which would be of great use to everyone including the *taupases* (artillery men) and Indians. They could thus learn English or French. Such a dictionary did not exist, and so the Catholic missionaries were using dictionaries written and printed by the Protestants. Dupuis said also that the Protestant pastors of Madras had made good use of the Catholic missionary dictionary manuscripts and they have repeatedly quoted the sources. They had also then brought out huge dictionaries and sold at 50 rupees per copy. According to Dupuis, the dictionary published was compiled by Johann Peter Rottler and it had been divided into four parts and had been defective, poorly composed, and words were fixed according to etymology and the alphabetic order was not properly respected. To find a single word one had to search at least in four or five places and this became tedious for the users.[5]

L.S. Dupuis noted that an excellent dictionary could be published therefore and it could be in four languages. This Latin-French-Tamil-English dictionary could be created and printed (in the format of the existing Latin–French dictionary) and this dictionary could have thousand pages approximately. This could be the most popular dictionary and in his opinion if 500 copies were printed, the dictionary could cost 3 rupees, and inclusive of all expenses it would amount to only 1200 rupees or at the most to 1500 rupees. This is inexpensive and even if it was sold at 10 rupees the dictionary would sell very well.

Dupuis noted sadly that his project had been unfortunately turned down and not approved by the Bishop of Pondicherry and the missionaries. Some clergy members had asked as to why a dictionary needed to be prepared in so many languages, and they also pointed out that the expenditure involved would be too high. Some others found that the dictionary work would be too voluminous in nature and unwieldy. Yet others felt that this dictionary would be of no use to Catholic Missionaries who worked in the interior lands and they also had no time to use and consult such a dictionary for want of time. As a result, Dupuis narrated that he had to drop this project as there were too many objections. He reported that he spent three months time in doing this work and had spent his precious time and the whole exercise had become useless. Thus he felt sad for wasting his time and energy. He finally added that he would be less taxed hereafter and so he could recoup his health since the project had been totally wound up.

Compiling Latin-French-Tamil Dictionary by Dupuis and Mousset for the use of the Seminary in Pondicherry and Printing in 1846

Clement Bonnand the Bishop of Pondicherry persuaded L.S. Dupuis to omit only English from his dictionary project so that the work could be manageable in trilingual and the same would be of great utility to the French missionaries in Pondicherry mission. The Bishop in his letter dated 15 April 1842 to Jean Tesson in Paris had mentioned that the priority had been assigned to the trilingual dictionary of Latin-French-Tamil and not to former idea to have Latin-French-Tamil-English. Dupuis had also accepted and acted according to the wishes of the Bishop. He stated that Dupuis had simply obeyed and commenced the work of compiling the *Dictionarium Latino-Gallico-Tamulicum* (Latin-French-Tamil dictionary). Louis-Marie Mousset another MEP missionary had also been persuaded and he expressed his willingness to join in this dictionary project and the work had commenced and started progressing. Mousset in his letter dated 19 April 1845 to Jean Tesson stated that he had been busily engaged in the work of Latin-French-Tamil dictionary which would be completed within some months.

The Latin-French-Tamil dictionary compiled by Dupuis and Mousset in Pondicherry finally got printed at the Catholic press in Pondicherry in 1846 where Dupuis was the manger of the printing press. The dictionary had the total of xviii+1392 pages.[6] We find mention of completion of the entire work of Latin-French-Tamil dictionary by Mousset vide his letter dated 10 February 1846 to Jean Tesson in Paris.[7] He informed that Dupuis had completed the printing of the dictionary which had 1410 pages. The dictionary printed in 1846 was also referred to by Dupuis in his letter dated

5 February 1847 to Jean Tesson while sending the copies of Latin-French-Tamil dictionary to Paris.[8]

In the preface of the dictionary Dupuis mentions Pierre Magny (1748-1822) the MEP missionary who started his work as a professor at the seminary and became its director during 1775-78 and later became also the superior of the seminary during 1778-81. His experience as teacher helped to prepare the *dictionnaire Latin-Tamoule* (Latin-Tamil Dictionary) and the work was never published until Magny's death on 1 May 1822. It was found useful for his compilation of the Latin-French-Tamil dictionary.[9]

Dupuis and Mousset had completed the compiling of the dictionary with the help of earlier existing materials. Publications were an arduous labour and Dupuis had said that they faced some obstacle but had the needed patience to withstand all the difficulties. He said that the senior missionaries had left good written material and that were to be printed or used for making new books. Many such writings had existed in huge quantity and these works were also without the names of the author priests. Dupuis said that he had appealed for the collection of such manuscripts to be published as books in the Catholic press in Pondicherry.[10]

Mousset had also announced once again vide his letter dated 10 November 1850 that the Latin-French-Tamil dictionary printed had 1410 pages.[11] The trilingual dictionary preparation and printing had thus taken a long time i.e. nearly four years. Dupuis and Mousset prepared the materials exclusively for teaching in the seminary in Pondicherry. In fact Latin was an important subject to be taught for ecclesiastical studies to the seminarians. French was needed for correspondence among the missionaries and the Tamils who aspired to be priests. Therefore they prepared and printed both grammar books and dictionaries targeting the seminary audience.

In the printed dictionary we find mention of the title page in three languages such as Latin, French and Tamil. We notice Latin words are given at first and then the meanings in French followed by Tamil. Abbreviations are given in the beginning for the use of the dictionary and the list in Latin and French is as follows:

Abbreviations in Latin

Abl. Ablatius

Acc. Accusativus

Act. Active

Adj. Adjective

Adv. Adverbium
Comp. Comparativus
Con. Conjunctio
Dat. Datius
Dim. Dimnutivum
F. Femininum
Freq. Frequentativum
Fut. Futurum
Gen. Genitivus
Hin. Hinc
I.e. Id est
Ind. Indeclinabile
Inf. Infinitivus
Int. Interjectio
Inus. Inusitatum
Ir. Irregulare
Loc. Locutio
M. Masculinum
N. Neutrum
Neg. Negatio
Nom. Nomen
O.g. Omnis generis
P. Pro
Part. Participum
Pass. Passive
Pl. Plural
Praep. Praepositio
Praes. Praesens
Praet. Praeteritum
Pro. Pronomen
S. Singularis
Subj. Subjunctivus
Subst. Substantium
Sup. Superlativus
V. Vide
V.a. Verbum activum
V.def. Verbum defectium

V.d. Verbum deponens
V.i. Verbum impersonale
V.n. Verbum neutrum
V.p. Verbum passivum
Vel. Vel
Voc. Vocativus

Abbreviations in French

Act. Action
Arch. Architecture
Esp. Espece Fig.de rhet.
Figura rhetorique

The dictionary begins with 'A' in Latin and then 'a' in French and the 'a' in Tamil giving the meaning in Tamil 'Muthal Yezhuththu'. The Dictionary ends with Zythum and giving the meaning in Tamil 'Vaarkothumbaikku kondu seyyum bhaanam'. Dupuis had also sent one copy of the Latin-French-Tamil dictionary to M. Base, the member of the institute and president of L'Ecole des Langues Orientales Vivantes. Another copy was sent to the Conservateurs de la Bibliothèque Imperiale in Paris. The MEP superiors in Paris were very glad to distribute it to the libraries and individuals in France. Mr. M. Base appreciated the printing of the dictionary and he sent a letter of praise on 29 May 1853 to Jean Tesson and thanked him for the dictionary copy sent. According to records, the price of the tri-lingual dictionary (that had 1410 pages) in 1853 in Pondicherry was five rupees. The selling price for outstations varied and it was fixed at five rupees and six annas.[12] In a book printed titled 'Anda Vizhlakka Sasthirathin Sangshebum Koodiya Bhumisasthira Noolaathaaram' in 1864 the price of the Latin-French-Tamil dictionary had been reported and it was sold at 4 rupees. This shows that the selling price was reduced and fixed again to promote local sales.

Preparing French-Tamil Vocabulary by Dupuis and Mousset and Printing in 1850

The MEP missionaries in Pondicherry were interested in the compilation and printing of *Vocabulaire Français-Tamoul* (French–Tamil vocabulary). Bonnand the Bishop of Pondicherry in his letter dated 4 October 1849 to the MEP Directors in Paris mentioned that the printing of the French-Tamil vocabulary would soon be completed. He also appreciated the efforts of Dupuis and Mousset for compiling successfully the vocabulary.[13]

L.S. Dupuis in his letter dated 6 June 1850 to Jean Tesson in Paris mentioned that he would finish soon printing the French-Tamil vocabulary within a few months.[14] Mousset in his letter of 10 November 1850 to Jean Tesson mentioned that the printing of the dictionary was over and the dictionary had 420 pages. He also mentioned that there were errors and corrections which had been indicated.[15] This French-Tamil vocabulary was originally intended as an appendix to the earlier trilingual dictionary but expanded into a larger work with a great number of words, which, although not in frequent use, sometimes caused difficulty to readers and translators as the compilers have confessed.[16]

The French-Tamil vocabulary thus compiled by Dupuis and Mousset was printed at Pondicherry in 1850. The book had a total of 420 pages.[17] Dupuis in his letter dated 6 June 1850 to Jean Tesson wrote that the grand dictionary of Latin-French-Tamil dictionary had greatly enabled him in the making of the bilingual vocabulary of French and Tamil.[18] He was easily able to reduce it though it took time. In the preface of the French-Tamil vocabulary it had been mentioned that the work contained 54,000 entries where the meanings are explained. It has been noticed that Mousset and Dupuis in the *Vocabulaire Française-Tamoule* had given an explicit Christian orientation with lot of words.

Dupuis and Mousset felt the need to use the earlier manuscript dictionaries circulated among the missionaries. They desired to add more to the existing materials which were prepared long ago. Dominique de Valence the Capuchin missionary in Pondicherry wrote a French-Tamil dictionary and the copies of his manuscript *Dictionnaire Français-Tamoule* (French-Tamil Dictionary) had survived at various repositories in Europe too.[19] His name and the year 1743 is mentioned in the dictionary. Valence had mentioned that it was for his own use and for the use of those who wish to commit themselves to the same apostolic ministry, that he undertook the compilation of French-Tamil dictionary. He said therefore that he dedicated the work for the greater glory of God the almighty.[20] Valence by following the normal alphabetical pattern then in use in the European dictionaries had begun to adopt the same pattern in the compilation of his dictionary. Valence wanted to make the reader to understand clearly the plan of his dictionary. He said that the alphabetical arrangement of words was followed without repetition. In the dictionary he had given as much synonyms as possible and explained the grammatical rules. He also said that he had used the dictionaries compiled by the earlier authors by adding more and more words. He had also omitted some entries. Valence said that his French-Tamil dictionary had been compiled for the use of French speaking people. It would be interesting to learn that how far the French missionaries clubbed and mixed the Franco-Tamil lexical traditions.

Compiling Tamil-French Dictionary of Volume I by Dupuis and Mousset and Printing in 1855

L.S. Dupuis and Mousset continued their efforts in compiling a Tamil-French dictionary. Dupuis in his letter dated 5 January 1854 to Jean Tesson mentioned that the *Dictionnaire Tamoul-Français* (Tamil-French dictionary) work had been satisfactorily progressing and it would be soon printed.[21] Mousset in his letter dated 3 August 1854 had said that the printing works of Tamil-French dictionary had commenced and 160 pages had been printed already.[22] He mentioned that the letter 'A' in Tamil had 133 pages. Mousset in his another letter dated 10 November 1854 to Jean Tesson wrote that printing of the Tamil-French dictionary had been well progressing.[23] Finally the Tamil-French Dictionary of volume one was printed in 1855 at Pondicherry.[24]

Dupuis in his letter dated 5 May 1855 informed Jean Tesson about the preface of the Tamil-French dictionary that saw the light of printing completed.[25] He also said the preface had 23 pages and the introduction had 9 pages. L.S. Dupuis desired that a printed copy of the Tamil-French dictionary to be displayed at the book exhibition to be held in Paris. Mousset in his letter dated 25 October 1855 to Tesson wrote that the Tamil-French dictionary printing had gradually progressed little by little.[26] Mousset even after four years said (vide his letter dated 10 January 1859 to Jean Tesson in Paris) that the first volume of Tamil-French dictionary had a total of xxxi+932 pages printed and it had alphabets commencing from A, Aa, to Gni and Gney in Tamil.[27]

It is important to note that Dupuis had clearly mentioned that he used the dictionary of Louis-Noel de Bourzés of 1724 and it helped him greatly in printing Tamil-French dictionary.[28] The French missionaries who arrived on the Tamil coast in the eighteenth century found the necessity to communicate well in Tamil and improve their skill in Tamil language. So they felt the need to compile a Tamil-French dictionary for their use. Their work also took several years for compiling because such a dictionary compilation needed a good command and mastery over Tamil and it was not that easy to learn in Pondicherry in the absence of learning tools. The missionaries noticed the bilingual dictionaries prepared by the Portuguese and also they used them. However, they wanted exclusively to compile a Tamil-French dictionary which would facilitate the French speakers in France and Pondicherry.

Using the Tamil-French Dictionary of 1724 of Louis-Noel de Bourzés by Dupuis and Mousset while Compiling Tamil-French Dictionary of Volume I and Printing in 1855

Louis-Noel de Bourzés the Jesuit had ventured to compile a *Dictionnaire Tamoul-Français*, in 1721 and he successfully completed it in 1724. It served as a working manual for the French missionaries. Bourzés says that he had consulted and used the Tamil-Portuguese dictionary of Antão Proença and that had been very useful for his work. This is found mention in his preface.[29] This proves that he was very much inspired by the trend set by Antão de Proença and enthused to compile the pilot dictionary of Tamil-French.[30] It may be mentioned that this dictionary of Bourzés had many obscure words in French. There are 13752 head-words often accompanied by examples of contextual use. Bourzés did not hesitate to admit where he was in doubt or cannot offer his readers any help.[31] He had openly mentioned the sources of dictionary compilation in his work. It could be gleaned from his entry at the dictionary which demonstrates his knowledge of the Tamil nikandus such as the agaraathi nikandu of 1594 AD by Revana Siddhar. In his preface he had also admitted that he used a French dictionary (author / compiler unknown) to help him in the compilation of Tamil-French dictionary.

Bourzés neatly said that a complete dictionary compilation demanded more ability and leisure. Yet he decided to compose and offer only a pilot dictionary. According to him those who use it should find the dictionary very good as to add whatever they may find lacking and also to remove whatever they might find in excess, besides free to correct the errors in it.[32] He said that there would be no point in searching in the dictionary for terms particularly related to the things unknown in India, for example the European sciences, or arts or trees or fish and the things peculiar to India which they did not have in Europe. He stated that he had placed under their generic terms in French such as, an (year) arbre (tree) arbrisseau (bush). He added that he had also placed some among the French terms.[33] He said that he had not entirely neglected senthamizh.[34] Words like karuman meaning blacksmith was unknown in Madurai region while the same term was found very common in Pondicherry area.[35]

Bourzés added that in his Tamil-French dictionary one would find a number of terms of grammar and prosody which were not to be found in the Tamil-Portuguese dictionaries compiled by the earlier missionaries.[36] He said that to the head-word he often added another, not to explain the French term, but to determine the sense of the Tamil term which followed as it was fairly rare to find Tamil terms which corresponded exactly to French. For the same reason he had used some Latin terms, either for the sake of economy, or because they seemed to him to correspond better to the Tamil, or because no other French word came to his mind.[37]

Bourzés said that his Tamil-French dictionary was meant for the use of the Jesuit priests in France who would come newly to Tamil country for evangelization work.[38] He mentions that he had extensively used Tamil-Portuguese dictionary of João de Sequeira while compiling his Tamil-French dictionary. Several remarks had been made by Bourzés on the usage of Tamil-French dictionary and he had given under eleven items. He said (under no.7) that he collected several terms from many places wherever he had lived in Tamil country.[39] He also said (under no.10) that he often omitted the participles or simply said they could be formed from the verbs according to the rules of the grammar.[40] He mentioned (under no.11) that as the past participle was formed from the past active, when he derived this participle from the past it was unnecessary to indicate the past active.[41] He also said that one would find at certain places in this dictionary the letter 'R'. or 'Rob' and this was to indicate that the term or expression was due to Roberto Nobili (1577-1656) the Jesuit. The letter 'F' indicated that the expression was found in the works of Andre Freyre (1625-1692) the Portuguese Jesuit whose Tamil was so correct that on hearing him, one could not distinguish between him and a native speaker. The letter 'l' or 'lus' indicated the expression from one or other of Portuguese dictionaries. The letter 'C' indicated terms taken from João de Sequeira's Tamil-Portuguese dictionary. The letter 'I' indicated that he had taken the word from one of the native dictionaries.[42] Finally Bourzés mentioned that the Tamil-French dictionary work was not completed and while he was correcting the errors in the dictionary, a request was made by his Jesuit superior to send a copy of the same and so it was sent from Pondicherry. Thus the dictionary was still in progress and he had sent it as it is.[43]

The dictionary of Bourzés contains firstly terms found in Tamil-Portuguese dictionaries except for some which Bourzés had rejected as inaccurate or in very vulgar usage and others which could be easily formed by the morphological rules. He said that one would find in this Tamil-French dictionary those terms contained in the five local dictionaries. The five dictionaries included Dhivaagaram the oldest, (middle of eighth century) second the Chudamani nikandu (1520 AD) the best known, third urisol the shortest, fourth Pingalam, (appeared between eighth and thirteenth century) the longest and the fifth Agaraathi, (1594 AD) the least accurate.[44] He said also that the variations in spellings had obliged him to often enter the same term at three or four different places. It was useful to have a work where one could find the meaning of an unfamiliar word whenever one came across it in speech or in reading.[45] According to Bourzés, Antão Proença the author of Tamil-Portuguese dictionary had rejected the poetic terms because they were rarely used.[46] Bourzés refers to the then existing Tamil-Portuguese dictionaries and regrets that when he began learning Tamil he did not have a dictionary to which he could add items, correct what he found also wrong. Bourzés was the first European to use the Tamil alphabetical order for head-words. It is interesting to find another copy of Tamil-French dictionary of Bourzés which was finally sent after completion of the work to Europe.[47] We find a copy of Tamil-French Dictionary of Bourzés in two volumes made available in 1750 for use.[48] His work was very popular in Pondicherry and we find corroborating evidence by Frei Paolino de San Bartolomeo. He said that during his stay for thirteen years from 1776 to 1789 he found several copies of Bourzés dictionaries that were made by individuals and they were widely in use.[49]

Dupuis and Mousset Using Beschi's Chaturagaraathi of 1732 Printed at Madras in 1824 and Reprinted in 1835 and his Tamil-French Manuscript Dictionary of 1744

Dupuis and Mousset used Beschi's Chaturagaraathi of 1732 and his Tamil-French manuscript dictionary of 1744. It is known that Beschi was perturbed over Louis-Noel de Bourzés' juxtaposition of 'H' and 'L' Tamil in one compilation and he gave his rationale for separating the two. Beschi had acknowledged that he used the Tamil-French dictionary of Bourzés while making his Tamil-French dictionary. Beschi in his work wished to explain ordinary Tamil, with a preface in which there are several rules one needs to know to be able to use the dictionary.[50] He commenced the work in 1744 and completed it in the same year. Being an ardent Jesuit he said that he had composed it for the greater glory of god, the honour of Mary, the virgin mother and the growth of the Roman Catholic Christian religion.[51] In his *avis au lecteur* (notice to the reader) of his Tamil-French dictionary he mentioned that the work had demanded a lot of time and energy and it had given him a great deal of trouble too, and there was nothing beautiful or attractive about it. He stated that his role was simply to give them the words. He added that he had played only a minor role in this work.[52]

Beschi said in the preface of his Tamil-French dictionary that he had been living in the hinterland of the Tamil coast for thirty years (besides three years since after arrival in Goa in 1711) where pure Tamil was spoken, uncorrupted by the influences which have appeared in some coastal areas through trading activities of the Europeans. In obedience to his superiors he took up the task of this Tamil-French dictionary.[53] We find copies of the dictionary mentioning the year 1744.[54]

According to Beschi, there were two dialects in Tamil, the kodunthamizh, the ordinary and the senthamizah or the elevated. Between these two dialects there was as much difference as there was between Latin and Portuguese. Having written already a dictionary (Portuguese-Latin-Tamil Dictionary) and a Synonymy (Chathuragaraathi) of the elevated dialect alone, which he explained via the ordinary, just as someone might use Portuguese to explain Latin words, the Tamil-French dictionary dealt with only with ordinary dialect and contained only those terms which were in common use either in speech or writing. He had also added certain Sanskrit words, whose use was accepted among the Tamils. These words could be easily distinguished from others by the letter 'G' used as label. If any of these Sanskrit words were to be pronounced in a way which cannot be indicated by the Tamil letters, he added the romanization. In this Tamil-French dictionary he had put only those Tamil words which were well attested. Words not in use, or used only by few people or only in a few places or rarely, were not included.[55]

Beschi had attempted to follow the order of the Tamil alphabet. Among the lexicographers of Hebrew and Greek, there was no one who did not follow the order of the respective alphabet. It was not difficult to accustom oneself to the Tamil alphabetical order. He cited the twelve vowels, (besides the two diphthongs ai and au), then the eighteen consonants divided into three classes, the six hard/rough, the six nasal/smooth and the six medial, part rough, part smooth and at the end certain Sanskrit letters with their shapes and sound at the end of the preface.[56]

From time to time he cited certain examples of Tamil sentences in order to help the reader better understand both the sense of the particular Tamil word, as well as the phraseology of the language. He mentioned that occasionally there are certain literary terms which were nevertheless used by all and it was useful to know these proverbs.[57] He indicated always the roots of the words using an 'R' as a label immediately preceding the radical. Thus one can learn to spell perfectly, and also to know thoroughly the sense of the Tamil word.[58] He had noted the general rules for adjectives obtaining in the literary language since several of these rules were also observed in the ordinary language, and through them one could knew the adjectival roots.[59] He mentioned the important rule in the Tamil literary language and the rule governed what was called vinai thogai in Tamil, an apocope by which one dropped the final letter of the future participle, with the sense of present, past and future time, depending on the context. If one could learn this rule well, the roots and meanings of several words would become obvious.[60]

In the Tamil-French dictionary of Beschi, the 'command' verbs as he had called them in his grammar work would not be found. He had left almost all of them because according to the rules of formation of these verbs were given in his grammar and it was easy to reconstruct the simple forms from which they were derived.[61] Beschi said that Tamil spellings could be ambiguous and so to avoid inconvenience he had thought to use certain signs which had been approved by several scholars of distinction and he gave these details too.[62]

The view of Innasi that Beschi did not compose Tamil-French dictionary needs to be reconsidered. Beschi himself mentions in the Latin preface of Chathuragaraathi 'Vulgaris language lexicon, ubi quaslibet lections latine, gallice ac Lusitane explico'. This proves that his work of Tamil-French dictionary was in progress.

The dictionaries of the French missionaries exhibited wealth of Tamil words. What was spoken and uncommon in literature was included as words in dictionaries. Meanings of words were given for the first time in prose and not in poetry form. Roots of verbs were brought separately and noted and the missionaries differentiated them from other words. In fact they gave an opportunity to learn equivalent French words for Tamil words. The colours of the apostolic artists in Tamil language and linguistics must be highly appreciated. The Tamil-French dictionaries were prepared with a practical approach to provide a handy dictionary to French missionaries learning Tamil for purposes of ordinary and evangelical work. On meeting a Tamil word a missionary could immediately look it up according to the relation of sound and symbol without having to learn the Tamil graphemic order. Thus this dictionary was for non-native learners constructed according to the users' needs. In some manuscripts of Tamil-French dictionary preserved in Paris no identification of the author could be made.[63] In the case of two manuscripts the years have been mentioned but the names of the authors could not be found.[64]

It is interesting to note that Muthu Vijaya Thiruvengadam Pillai in his Tamil diary refers to purchase of the Tamil-French dictionary by him on payment of money on 15 May 1795.[65] Again on 19 May 1795 he had also purchased a French dictionary after paying 30 rupees.[66] Thus natives in Pondicherry have widely used the dictionaries for their use. It is no surprise therefore that Dupuis and Mousset had used Beschi's Chaturagaraathi of

1732 printed at Madras in 1824 and reprinted in 1835 and Beschi's Tamil-French manuscript dictionary of 1744.

Mousset and Dupuis in the Tamil-French dictionary compiled under the word 'prabhandam' has listed the 96 types of Tamil poetry and the details of composition in French. A note at the end of the term had been also given in French. Beschi in his Chaturagaraathi had earlier mentioned the same 96 prabandhams and gave also the details of each genre along with the details in Tamil. Dupuis who had used Beschi's Chaturagaraathi had used the data in his compilation but wrote in French.

1. Saathakam	2. Pillaikavi
3. Bharani	4. Kalambagam
5. Akaporutkovai	6. Ayinthinaicheyyul
7. Varukkakkovai	8. Mummanikkovai
9. Angamaalai	10. Ashtamangkalam
11. Anuraagamaalai	12. Rettaimanimaalai
13. Inamanimaalai	14. Navamanimaalai
15. Naanmanimaalai	16. Naamamaalai
17. Palasanthamaalai	18. Panmanimaalai
19. Manimaalai	20. Pugalchimaalai
21. Perumagilchimaalai	22. Varukkamaalai
23. Meykirthimaalai	24. Kaapumaalai
25. Veninna maalai	26. Vasanthamaalai
27. Thaaragaimaalai	28. Urpavamaalai
29. Thaanaimaalai	30. Mummanimaalai
31. Thandagamaalai	32. Veeravetchimaalai
33. Vetrikarandhaimanjari	34. Porkezhuvanji
35. Varalaattruvanji	36. Serukkazhlavanji
37. Kaanchimaalai	38. Nochimaalai
39. Uzhinzhaimaalai	40. Thumbaimaalai
41. Vaagaimaalai	42. Vaathoranamanjari
43. Yencheyyul	44. Thogainilaicheyyul
45. Oliandhaathi	46. Pathithandhaathi
47. Nootrandhaathi	48. Ulaa
49. Ulaamadal	50. Vazhamadal
51. Orupaaorupaththu	52. Irupaairupadhu
53. Aattrupadai	54. Kanpadainilai
55. Thuyiledainilai	56. Peyarinnisai
57. Urinnisai	58. Peyarnerisai

59. Urnerisai
60. Urvenpaa
61. Vilakkunilai
62. Puranilai
63. Kadainilai
64. Kaiyarunilai
65. Thasaangapaththu
66. Thasaangathayal
67. Arasanviruththam
68. Nayanapaththu
69. Payotharapaththu
70. Pathaathikesam
71. Kesaathipaatham
72. Alangarapanchagam
73. Kaikilai
74. Mangkalavazhllai
75. Thoothu
76. Naarpathu
77. Kuzhamagal
78. Thaandagam
79. Pathigam
80. Sathagam
81. Seviarivuroovu
82. Vaayaurai vaazhthu
83. Puranilai vaazhthu
84. Pavanikaathal
85. Kuraththipaattu
86. Uzhathipaattu
87. Voosal
88. Yezhukoottirukkai
89. Kadigaivenpaa
90. Sinnappoo
91. Viruththavilakkanam
92. Muthukaanchi
93. Iyanmozhivaazththu
94. Perumangkalam
95. Perungkaapiyam
96. Kaapiyam

N:B *Notre intention, en faisant connaitre succinctment ces diverses productions de la poesie indeinne, a ete de donner quelque idée de ce qui fait en grande partie le fond et l'objet de la literature de ce pays. Si dans ce nombre d'ourages nous avons parle aussi de ceux qui sont erotiques, c'est afin qu'on puisse le eviter et les defender plus facilement.*

Dupuis and Mousset Using the Dictionary of Johann Peter Rottler (1834), Dictionary of Joseph Knight and Levi Spaulding Printed in Jaffna (1842), and Tamil-Latin dictionary of Joseph Gury Printed at Nagapattinam (1852)

Dupuis and Mousset stated that since the manuscripts were very incomplete, the help the purely Tamil, or Tamil-English dictionaries afforded them still left them a huge task to complete. Nevertheless, they said that they were pleased to offer their gratitude for the principal sources they have mined, which included Johann Peter Rottler's 'A Dictionary of the Tamil and English Languages', printed at Vepery during 1834-41, the work of Joseph Knight and Levi Spaulding, 'A Manual Dictionary of the Tamil Language', printed at Manipay, in Sri Lanka in 1842 and of the same authors another work titled 'An English and Tamil Dictionary or Manual Lexicon for schools, Giving in Tamil all important English words and the use of many phrases', printed at Jaffna in 1842, and the lithographed Tamil-Latin dictionary of Joseph Gury, the Jesuit titled '*Dictionarium Tamulico-Latinum ex Dictionariis Beschi et Rottler Conflatum*' printed at St. Joseph's College Press in Nagapattinam in 1852. Thus it is interesting to note that they were very up-to-date and had access to all these dictionaries.[67]

Dupuis and Mousset mentioned that the Jaffna dictionary did not always agreed with Rottler's especially with regard to spelling. Without wishing to judge between them, they had generally followed the latter. However, they wished to warn the readers that where the author contradicts himself on certain points, where usage appears not to be well established, they have not presumed to make absolute pronouncements either. Thus, after a noun ending in vu, preceded by a double rough consonants such as ka, sa, da, tha, pa, ra, and followed by a verb beginning in ka, sa, tha, or pa, he sometimes doubles the initial and sometimes not. He had put for example, uppu paarkirathu, uppu podukirathu, and mannippu kaetkirathu, mannippu pannukirathu. It seemed to Mousset and Dupuis, however, that in such a case, not doubling these consonants was becoming the more general usage.[68]

The Preface to the Tamil-French Dictionary of 1855 by Dupuis and Mousset

Dupuis and Mousset in the preface to the Tamil-French dictionary of 1855 mentioned that since divine providence designed then to send evangelical workers in greater numbers to India than before, the desire to obtain for them the means of learning perfectly and easily the Tamil language of the people whom they have come to enlighten with the flame of faith, inspired to them to take up the editing and printing of this Tamil-French dictionary. Therefore they wished to offer it to the public for the first time. They said that an enterprise doubtless beyond their capacity, but whose necessity was so greatly felt that, within the limits of their ability, they should try to respond to the general expectations and fill the gap left by their Latin-French-Tamil dictionary printed in 1846 and their French-Tamil vocabulary published earlier in 1850.[69]

While attaining their first aim, they would be equally happy if, through this dictionary work they could prove themselves useful to the French colony of Pondicherry, where the study of French and Tamil had for some years had developed favourably, thanks to the encouragement of the government. The enthusiastic young students of Pondicherry would, they hope, find in this book a powerful aid to their studies. They promise themselves that the usefulness of this Tamil-French dictionary would not be limited only to the French settlements.[70]

In order to make it useful to all, Dupuis and Mousset had brought together, in alphabetical order, as large a number of words from ordinary and poetic Tamil as they had been able. These two varieties of the same language, they would say almost in general distinct, did not have demarcations absolute or clear-cut enough to be able to separate them in a dictionary without leading whoever studies either one into frequent difficulty. In the way of all lexicographers, they had been able, and indeed had needed, to benefit from the compilations of all their predecessors in this profession; all the more that, among them, the foreigners had already used the valuable works of the missionary priests of the former times, who were the first to compile real dictionaries of Tamil. Dupuis and Mousset said they are heirs to their work. Moreover, as up-to-now no Tamil-French dictionary had been published.[71] It must be mentioned that they were unaware of the Agaraathi Thamizh i. e., dictionary printed in Paris already by André Blin in 1831.

So as to make known, as far as possible, the mythology, religion, usages, sciences, teachings, proverbs and other such information, as well as the language, of the people, Dupuis and Mousset had sometimes added to the translation of the words some short explanation or other useful details they had been able to find. They said that let one imagine that they believe, or proclaim as true, what they included in this Tamil-French dictionary concerning the mythology of the Indians or their superstitious practices. They felt that a dictionary should give the sense of words, without thereby authorizing their abuse or the false teachings for which they may be used. On many occasions, they have formally expressed their disbelief in, or aversion for, the false maxims, the absurdities and the depravities of Hinduism. If in order not to repeat themselves, they have not always made their feelings obvious, and so one should not draw the conclusion that they admit in the least the errors of superstitions, which they desire from the bottom of their hearts to see this country renounce, so as to embrace the truth and enter into the way of salvation and happiness.

To facilitate the in-depth study and analysis of this Tamil language, Dupuis and Mousset said that they had put in parentheses the origin of nearly every word. Just this will very often suffice to show up the absurdity and abomination of Hindu theology. Thus Aanaimurugan and Aguvaagana were the names of Vinayaka, or Pillayar, one of the most popular divinities. The first meant the one who had the face or head of an elephant, the second, 'the one who rode a large rat'. Indeed the Hindus believed that the son of Siva, having had his head accidentally crushed-the cause of which was variously reported-someone cut off the head of an elephant and fixed it onto his body, where it fitted well enough to replace his own. And that elephant head, pot-bellied god has nothing more than a large rat as a mount. Harihara Puthiran meant Ayyapan the son of Vishnu and Siva. The Hindus claim that Siva had him through Vishnu, who had been metamorphosed into an enchantress, a fable which alone sufficed to characterize the morals of the greatest of the gods, and the religion of India. Kama was the name of the god of impurity, and meant the lustful one. That of Rathi or Irathi his wife meant lust or copulation and was used to denote the most obscene things. Dupuis and Mousset in their preface said that this Tamil-French dictionary contained many such examples. They hoped that these etymologies could indeed serve powerfully to resolve these two important questions. Is Tamil derived from Sanskrit? Is it the mother–language?[72]

As compilers of the dictionary they added that there was no doubt that Tamil has borrowed a host of expressions from Sanskrit, or Grantha, especially the vocabulary of mythology and science. Their etymologies and their use of asterisk indicated this at first glance. But most of these words were known only to scholars and were alien to ordinary Tamil. The language of the people generally consisted of words of pure Tamil origin, for which Sanskrit offers no primitive forms.[73] Thus Dupuis and Mousset indicated Sanskrit words in the Tamil-French dictionary by an asterisk and also other informative entries on Indian culture with an important caveat.

Dupuis and Mousset mentioned that Tamil has borrowed few words from other languages other than Sanskrit, except for the language used in the various offices of civil, military and judicial administration, or by the different groups of workers who have contact with Europeans. In these occupations, a kind of usage has grown up from several different foreign languages, and which varied from place to place, therefore the words had not been analyzed nor did they appear in dictionaries.[74]

Preparing the Tamil-French Dictionary of Volume II by Dupuis and Mousset and Printing in 1862

Dupuis and Mousset had tirelessly continued their efforts in the making of Tamil-French dictionary. The work of the second volume had begun from the Tamil alphabets of Ma, Ya Va, etc. Dupuis in his letter of 10 November 1857 to Jean Tesson said that the ongoing second volume of Tamil-French dictionary work was progressing very well.[75] Mousset in his letter of 6 July 1859 to Jean Tesson said that 3/4th of the Tamil-French dictionary work had been completed.[76] Dupuis in his letter to Tesson on 23 September 1859 told that Tamil-French dictionary would soon be printed.[77]

Dupuis in his letter to Tesson dated 23 December 1859 speaks of the grand success of compilation of the Tamil-French dictionary.[78] Mousset in his letter of 24 December 1859 mentioned that Tamil-French dictionary compilation would be soon completed.[79] In his letter Mousset wrote on 3 September 1860 that the Tamil-French dictionary which was in making for a long period and it took around 7 years, 5 months and 7 days to finish. He reported that the dictionary would be printed within one year.[80] According to records we find that the second volume of the Tamil-French dictionary was printed in 1862. It had xx+1115 pages.[81]

Using the French-Tamil Manuscript Dictionary of 1745 of Beschi while Compiling French-Tamil Dictionary by Dupuis and Mousset and Printing in 1866

Dupuis and Mousset with their excellent command over French and Tamil had continued to work in making a good French-Tamil dictionary by improving upon their earlier published *Vocabulaire Français-Tamoul* of 1850. Mousset in his letter dated 25 November 1860 to Jean Tesson mentioned the progress in compilation of this French-Tamil dictionary.[82] They successfully completed the work and it was printed in 1866.[83]

The dictionary has ten parts, such as le texte etudie, les destinataires, les sources, le nomenclature, les articles, le richesse, l'originaliter des lexicographes, leur honnetete, les deficinces, l'influence du dictionnaire. The dictionary begins with the word 'A' in French and ends with word Zygomatique. In compilation of this Dupuis and Mousset has used the French-Tamil dictionary of Beschi of 1745.

Beschi had planned to compose little French-Tamil dictionary and he commenced the work in 1745 and successfully completed it in the same year. We find a note that this little French-Tamil dictionary was compiled by him which was enough to enable a person to talk about any subject.[84] A copy of the French-Tamil dictionary made in Pondicherry was sent to Paris in 1750, within five years and it is now preserved at the National Library.[85] A copy of the same *dictionnaire français-tamoul* made in 1845 in Pondicherry is also available and this copy was made by Jagannathan in 1849 and collated by Chinnadurai Athisayam in 1853.[86] It contains also the list of words in French and Tamil by topics like des elements, de l' air, du feu, de la terre etc. (vide folios. 117a-121b) and sixty years of Tamil calendar (vide folios. 122a-123a). Other details include such as Tamil declensions (vide folios 124a-125a), Tamil conjugations (vide folios 125b-40a) besides des annees et des mois indiens en faveur des missionnaires qui sont dans l'interieur des terres ou onne connait pas notre Kalandrier. Thus copies of French-Tamil dictionaries were sent from time to time in ships that sailed from Pondicherry to France.[87]

Anquetil Duperron the traveller during the period of stay in India (1758-1761) had mentioned that Fr Emmanuel, the superior of the Jesuit mission in Pondicherry had spent twenty rupees for making a copy of the French-Tamil dictionary composed by Beschi.[88] This copy of the French-Tamil Dictionary was carried by Anquetil Duperron in his bag to Paris. On his arrival on 14 March 1762 in Paris he kept it carefully with himself. Later it found way to the National Library, Paris.[89] Under these circumstances, based on the above evidences the views of Leon Besse and Julien Venson that Beschi did not compile the French-Tamil dictionary may summarily be rejected.

Reprinting French-Tamil Dictionary in 1870

The earlier printed dictionary of *Français-Tamoul* in 1850 and 1866 had sold well. There was a demand for the book when copies were unavailable. Therefore French-Tamil dictionary was reprinted by Dupuis at the Catholic press in 1870.[90] Finally the next edition also appeared when volume one and two were put together and the French-Tamil dictionary appeared in 1873. Fr Jules François Bergez had by that come to assist Fr L.S. Dupuis in the printing press. Dupuis had asked his superiors in Paris to appoint a missionary to succeed him at the printing press. Hence Bergez was appointed in 1871 and he assisted Fr. Dupuis in the press. Fr. Bergez also became the Rector of the major seminary at Pondicherry in 1872 and he took over the office of the manager of the printing press after the death of Dupuis in June 1874.

The Printed Dictionaries Sold in the Overseas

The French-Tamil dictionaries and Tamil-French dictionaries had attracted sales in the overseas. The French government had taken much interest in the diffusion.[91] Dupuis printed a catalogue of the books published which reveals a comprehensive list with titles, authors and prices and a brief description of the books to target the audience.[92] Enclosing such a catalogue of the works brought out at the Catholic press Dupuis in his letter to the Minister of Marine in Paris wrote in 1863 and suggested that the books printed in Pondicherry would be very useful to the European officers living in India and France and also for the immigrants and indentured labourers living in the overseas French plantations and colonies where the promotion of the knowledge of Catholicism was an undeniable good.[93] In order to let the Government in France to decide whether the books were worthy of purchase he enclosed a printed catalogue, which could be of great use in learning Tamil in the various French colonies around the world since a large population migrated spoke Tamil and the government officials could also use it. He stated that the books would be also of use for Tamil classes offered in Paris and would be greatly useful in teaching and inculcating morals to the large number of immigrants who work in the French colonies. He stated that if the Minister for Marine recommended the books for purchase the printing press in Pondicherry would get some encouragement, and it would contribute even more good to the public.

The French Governor of Guiana ordered a shipment of bilingual (Tamil-French) books to be given as annual prize to the children of the Tamil speaking indentured labourers attending the plantation schools in 1863. Later in 1865 a single shipment reached French Guiana to become among the first collections of bilingual books to be traded across in the far-flung peripheries of the French empire. At the request of Dupuis a book catalogue was forwarded by the French Governor of Pondicherry to all other French Governors in Reunion, Martinique, Guadeloupe and Cayenne. On Dupuis behalf, the Governor of Pondicherry also wrote to the Minister of the Navy and Colonies in Paris about the Catholic printing press. He mentioned about the Tamil-French dictionary and the grammar book that was published.[94] Although all the governors were contacted, the French governor of Guiana only responded and he ordered 76 volumes of books, including dictionaries and grammars.[95]

To conclude it may be mentioned that both the Tamil-French dictionary and the French-Tamil dictionary prepared by missionaries had remained in manuscript form for a very long time in Pondicherry. These dictionaries were chiefly written for the use of Europeans and served as learning tools. Dominique de Valence's Tamil-French dictionary was a reference work for missionaries who came across unfamiliar words in Tamil. Dictionaries of the other missionaries had helped in propagating Tamil knowledge for love of Jesus Christ too. The Tamil-French Dictionary in print appeared in Pondicherry in two volumes in 1855 and in 1862 chiefly through the sincere efforts of L.S. Dupuis and L.M. Mousset the priests of the société des missions étrangères de Paris. The compilers of the printed work had greatly benefitted from all the earlier manuscripts of dictionaries circulated among the users.

Notes

1. British Library, London, (hereafter BL) T 5908, *Agaraathi Thamzih: Dictionnaire Français-Tamoul et Tamoul-Français dedie a Moniseur Ministre de la Marine par A. Blin Lieutenant de Cypahis, Membre de al Société Asiatique de Paris*, Cher dendey Dupre Imp.lib. Rue Richelieu, 47 bis, a Paris.
2. Archives de la Société des Missions Etrangères de Paris (hereafter AMEP), Correspondance Pondichéry-Paris, Lettres, vol. 1000, pp. 660-1.
3. Thomas Anchukandam, *The First Synod of Pondicherry 1844: A Study Based on Archival Sources*, Bangalore, 1994, pp. 41-2, 78.
4. Ibid., pp. 73-4.
5. AMEP, Correspondance Pondichéry-Paris, Lettres, vol. 999, see the letter of Louis Savinien Dupuis dated 18 August 1841 to Bonnand the Bishop of Pondicherry.
6. *Dictionarium Latino-Gallico-Tamulicum: Dictionnaire Latin-Français-Tamoul, Latheen Thamizh Piranchu Agaraathi, Auctoribus duobus Missionariis Apostolicos Congregationis Missionum ad Exteros*, Puducherii, E Typographo Missionariorum Apostolicorum dictae Congregationis, 1846.
7. AMEP, Correspondance Pondichéry-Paris, Lettres, vol. 1000, see the letter of Louis-Marie Mousset dated 10 February 1846 to Jean Tesson in Paris.
8. AMEP, Correspondance Pondichéry-Paris, Lettres, vol. 1000, see the letter of Dupuis dated 5 February 1847 to Jean Tesson.
9. *Dictionarium Latino-Gallico-Tamulicum*, see the preface.
10. Adrien Launay, *Histoire des Missions de l'Inde: Pondichéry, Maissour, Coimbatour*, 4 vols., Paris, 1895-1898, vol. III, pp. 134-5.
11. AMEP, Correspondance Pondichéry-Paris, Lettres, vol. 1001, see the letter of Louis-Marie Mousset dated 10 November 1850.
12. Centre des Archives d' Outre–Mer, Aix–en–Provence (hereafter CAOM), Inde n.d, *Catalogue des livres de l'Imprimerie de la Maison des missions etrangeres a Pondichery, Circulaire addresse aux colonies et relative a une souscription a des ouvrages en langue tamoule edites par l'imprimerie de la mission Catholique en Inde*, 1863-1865; CAOM, Inde 1863, *Envoie du catalogue des ouvrages en langue tamoule edites par l'Imprimerie de la mission*.
13. AMEP, Correspondance Pondichéry-Paris, Lettres, vol. 1001, see the letter of Clement Bonnand the Bishop of Pondicherry dated 4 October 1849 to the MEP Directors in Paris.
14. AMEP, Correspondance Pondichéry-Paris, Lettres, vol. 1001, see the letter of Dupuis dated 6 June 1850 to Jean Tesson.
15. AMEP, Correspondance Pondichéry-Paris, Lettres, vol. 1001, see the letter of Louis-Marie Mousset dated 10 November 1850 to Jean Tesson.
16. *Vocabulaire Français-Tamoul, Compose par Deux Missionaires Apostoliques de la Congregation des Missions-Etrangères et enrichi d' utiles renovis a leur dictionnaire Latin-Français-Tamoul*, Pondichéry, Imprimerie des Missionnaires Apostoliques, 1850, p. 9. The text runs: 'un grand nombre de mots, qui, bien que peu usites, peuvent quelque fois embarrasser les lecteurs ou les traducteurs'.

17. *Vocabulaire Français-Tamoul, Compose par Deux Missionaires Apostoliques de la Congregation des Missions-Etrangères et enrichi d'utiles renovis a leur dictionnaire Latin-Français-Tamoul*, Pondichéry, Imprimerie des Missionnaires Apostoliques, 1850.
18. AMEP, Correspondance Pondichéry-Paris, Lettres, vol. 1001, see the letter of Dupuis dated 6 June 1850 to Jean Tesson.
19. Det Kongelige Bibliotek (hereafter KB), Kobenhavn, Oriental Department, MSS Cod Tamul 30 (UB); Bibliothèque Nationale, Paris (hereafter BNP), MSS Indien, no. 219, *Dictionnaire Français-Tamoul suivi de la Langue Malabare Vulgaire*.
20. BL, MSS EUR, E.121, *Dictionnaire et Grammaire Français-Tamoul*, see paragraph 1.
21. AMEP, Correspondance Pondichéry-Paris, Lettres, vol. 1001, see the letter of Dupuis dated 5 January 1854 to Jean Tesson.
22. AMEP, Correspondance Pondichéry-Paris, Lettres, vol. 1001, see the letter of Louis-Marie Mousset dated 3 August 1854.
23. AMEP, Correspondance Pondichéry-Paris, Lettres, vol. 1001, see the letter of Louis-Marie Mousset dated 10 November 1854 to Jean Tesson.
24. Bibliothèque Asiatique des Missions Etrangères de Paris (hereafter BAMEP), B 2304, *Dictionnaire Tamoul-Français, par Deux Missionnaires Apostoliques de la Congregation des Missions Etrangères: Anniya Desaththu Pothagar Sabaiilulla Apostholikku gurukkal iruvar seytharuliya Thamizh Piranchu Agaraathi*, Pondichéry, Imprimerie des Missionnaires Apostoliques de la dite Congregation, 1855.
25. AMEP, Correspondance Pondichéry-Paris, Lettres, vol. 1001, see the letter of Dupuis dated 5 May 1855 to Jean Tesson.
26. AMEP, Correspondance Pondichéry-Paris, Lettres, vol. 1001, see the letter Louis-Marie Mousset dated 25 October 1855 to Jean Tesson.
27. AMEP, Correspondance Pondichéry-Paris, Lettres, vol. 1002, see the letter Louis-Marie Mousset dated 10 January 1859 to Jean Tesson.
28. BAMEP, B 2304, *Dictionnaire Tamoul-Français*, 1855, p. vii.
29. BNP, MSS Indien, no. 213, pp. 1-2, *Dictionnaire Tamoul et Français*; see also, MSS Indien, no. 214; 'Description of Two Manuscripts in the Library of Madras Literary Society 2. De Bourges' Dictionnaire Tamul-Français (French Preface of De Bourges)', *Madras Journal of Literature and Science*, Third Series, 1864, pp. 111-15.
30. John Rylands University Library, Manchester (hereafter JRULM), Tamil Manuscripts, no. 6, *Essay du Dictionnaire Tamoul Français*; Government Oriental Manuscripts Library (hereafter GOML), Chennai, MSS D8 and MSS D9.
31. Gregory James, *Col- Porul: A History of Tamil Dictionaries*, CREA, Chennai, 2000, p. 117.
32. JRULM, Tamil Manuscripts, no. 6, preface, see paragraph 1.
33. Ibid., paragraph 2.
34. Ibid., paragraph 8.
35. Ibid.

36. Ibid., see paragraph 9.
37. Ibid., paragraph 11.
38. Archives Françaises de la Compagnie de Jésus (hereafter AFCJ), Vanves, Hauts-de-Seine, Fonds Brotier, vol. 86, fls.55-6, see the letter of Louis-Noel de Bourzes to Etienne Souciet in 1730; MSS vol. 86, fl. 65, see the letter of Louis-Noel de Bourzes to his brother dated 7 November 1735.
39. JRULM, Tamil Manuscripts, no. 6, preface, see paragraph 21.
40. Ibid., paragraph 24.
41. Ibid., paragraph 25.
42. Ibid., paragraph 27.
43. Ibid., paragraph 25.
44. BNP, MSS Indien, no. 213, *Dictionnaire Tamoul et Français* (avec des notes grammaticales) par le P. de Bourzes. This volume one contains 284+9 folios in two columns per page with entries in Tamil alphabets from A to Tha. MSS Indien, no. 214, *Dictionnaire Tamoul-Français* par le P. de Bourges. This second volume contains 263 +1 folios with entries in Tamil alphabets from thaa to sho. See also, MSS Indien, no. 208; no. 209, *Dictionnaire Tamoul-Français* par le. P. de Bourzes. The copy was made in 1750 and has 276+9 folios.
45. The two volumes of Dictionnaire Tamoul et Français, avec des notes grammaticales par le P. de Bourzes contain 1104 folios. BNP, MSS Indien, no. 213, fls.567; MSS. Indien, no. 214, fls. 537.
46. BNP, MSS Indien, no. 213.
47. JRULM, Tamil Manuscripts no. 5; Bibliotheque Royale Albertine, Brussles, Belgium, MS no. 21881, pp 1-531. It is in two columns and the sixty year Tamil calendar cycle of 1747-1807 is found.
48. Julien Vinson, *Catalogue des Manuscripts Tamouls*, Paris, 1867; Antoine Cabaton, *Catalogue Sommaire des Manuscripts Indiens, IndoChinois et Malayo-Polynesiens*, Paris, 1912; BNP, MSS Indien, no.188, 189, 208 and 209.
49. Paolonio da San Bartolomeo, *Voyage to the East Indies*, London, 1800, p. 192.
50. BNP, MS 215; BL, London, MS Or.1308, see *Dictionnaire Tamoul-Français pour expliquer le tamoul vulgaire avec une preface ou sont plusieurs regles qui l'faut scavoir avant de faire usage du dictionnaire, compose par le Reverend Pere Constant Joseph Beschi de la Compagnie de Jesus missionaire de l'Inde l'an 1744 Ad Majorem Dei Gloriam, Virginis Matris Maria de Jesus et religionis Christiana Catholica Romana Incrementum Amen*. BNP, MSS Indien, no. 217, *Dictionnaire Tamoul-Français*. This copy was made in 1750 and has 2+1+257 folios. BNP, MSS Indien, no. 210, *Dictionnaire Tamoul-Français.* This copy was made in 1760 and we find here the Tamil calendar which consists of sixty year cycle and the corresponding Christian years from 1730 to 1790. The manuscript has 275 +12 pages. BNP, MSS Indien, no.211, *Dictionnaire Malabare-Français*. The first volume has 157 pages and an introduction on sounds and syllables. BNP, MSS Indien, no. 212, *Dictionnaire Malabare-Français*. This is the second volume that begins from page 158 and ends at page 314.

51. BL, MSS Or.1308, fl.1a; The manuscript copy of Dictionnaire Tamoul-François abrege de celui du R. P. Beschi, is undated and contains 325 pages. BNP, MSS Indien, no.215 has 203 +1 folios and made in two columns. This copy of Dictionnaire Tamoul-Français was prepared by Muthusami and Jagannathan and they completed the work in 1849 for Edouard Ariel.
52. Institut Française de Pondichéry (hereafter IFP), MSS DICT-TA, 13.1, see paragraph 1.
53. Ibid., paragraph 2.
54. BL, MSS Or. 1308.
55. IFP, MSS DICT-TA, 13.1, see paragraph 3.
56. Ibid., paragraph 4.
57. Ibid., paragraph 7.
58. Ibid., paragraph 8.
59. Ibid., paragraph 12.
60. Ibid., paragraph 13.
61. Ibid., paragraph 14.
62. Ibid., paragraph 15.
63. BNP, MSS Indien, no. 211, fls.1-157; MSS Indien, no. 212, fls.158314.
64. BNP, MSS Indien, no. 217, fls.1-257, *Dictionnaire Tamoul-Française*, Precede d'un Syllabaire, AD 1750; BNP, MSS Indien, no. 210, fls.1-275, *Dictionnaire Tamoul-Française*, AD.1760.
65. BNP, MSS Indien, no. 157, fl. 50; S. Jeyaseela Stephen, *Muthu Vijaya Thiruvengadam Pillai Naatkurippu, 1794-1796*, Pondicherry, 1999, pp. 260-1.
66. S. Jeyaseela Stephen, Muthu Vijaya Thiruvengadam Pillai, p. 262.
67. BAMEP, B 2304, *Dictionnaire Tamoul- Français*, 1855, see the preface, p. vi.
68. Ibid.
69. Ibid., p. v.
70. Ibid.
71. Ibid., p. vi.
72. Ibid., p. vii.
73. Ibid.
74. Ibid., p. xiii.
75. AMEP, Correspondance Pondichéry-Paris, Lettres, vol. 1002, see the letter of Dupuis dated 10 November 1857 to Jean Tesson.
76. AMEP, Correspondance Pondichéry-Paris, Lettres, vol. 1002, see the letter of Louis-Marie Mousset dated 6 July 1859 to Jean Tesson.
77. AMEP, Correspondance Pondichéry-Paris, Lettres, vol. 1002, see the letter of Dupuis dated 23 September 1859 to Jean Tesson.
78. AMEP, Correspondance Pondichéry-Paris, Lettres, vol. 1002, see the letter of Dupuis dated 23 December 1859 to Jean Tesson.
79. AMEP, Correspondance Pondichéry-Paris, Lettres, vol. 1002, see the letter of Mousset dated 24 December 1859 to Jean Tesson.

80. AMEP, Correspondance Pondichéry-Paris, Lettres, vol.1002, see the letter of Louis-Marie Mousset dated 3 September 1860 to Jean Tesson.
81. BAMEP, B2304, *Dictionnaire Tamoul-Français: Thamizh Piranchu Agaraathi Rendam Pangu, par Deux Missionnaires Apostoliques de la Congregation des Missions-Etrangères*, Pondichéry, Imprimerie des Missionnaires de la dite Congregation, 1862.
82. AMEP, Correspondance Pondichéry-Paris, Lettres, vol.1002, see the letter of Mousset dated 25 November 1860 to Jean Tesson.
83. Bibliothèque Nationale de France (hereafter BNF), *Dictionnaire Français-Tamoul: Piranchu Thamizh Agaraathi*, Pondichéry, Imprimerie des Missionnaires de la dite Congregation, 1866.
84. BL, London, MS Or.1308; BNP, MSS Indien, no. 216; *Petit dictionnaire Français Tamoul compose par le Reverend Pere Constat Beschi, missionaire de la Compagnie de Iesu. Dans le Madure avec l'aide du Tamoul François suffisant pour pouvoir s'enoncer sur tout*.
85. BNP, MSS Indien, no. 218. This manuscript has 308 +1 folios.
86. BNP, MSS Indien, no. 216. This manuscript is 2 +145 folios with two columns per page.
87. BNP, MSS Indien, no. 215, *Dictionnaire Français et Tamoul*. This manuscript was received in 1845 and it has 203 pages. BNP, MSS Indien, no. 216 and it was copied by Jagannathan and Chinnadurai Athisayam and thoroughly checked and completed in 1853. BNP, MSS Indien, no. 220, *Dictionnaire Français et Tamoul*. This manuscript copy was made in 1850. The French entries were copied by Adolphe Pitatuan while the Tamil entries in the manuscript copy were made by Jagannathan.
88. Anquetil Duperron, *Relation de Voyage en Preliminiaire a la Production du Zend-Avesta*, eds., Jean Deloche, Manonmani Filliozat and Pierre Sylvain Filliozat, *Voyage en Inde*, 1754-1762, Paris, 1997, p. 238.
89. Anquetil Du Perron took also a copy of the manuscript of Tamil grammar in Latin composed by C.J. Beschi and a palm-leaf manuscript in Tamil which contained description of the temple of Chidambaram. Abbe Jean-Jacques Barthelemy procured for him a pension, with the appointment of interpreter of oriental languages at the Royal Library in Paris. Anquetil Du Perron deposited his 180 manuscripts to La Bibliothèque du Roi in 1762. He was elected as associate of the Academy of Inscriptions in 1763 and began to arrange for the publication of the materials he had collected during his travels. Anquetil Du Perron published a list of his own oriental manuscripts 'apportes de l'Inde' in appendix III of his Zend Avasta. See, Jean-Marie Lafont, *Indika: Essays in Indo-French Relations, 1630-1976*, Delhi, 2000, pp. 113-4.
90. Navis Sybil Abarna Roy, *La Contribution de Louis Savinien Dupuis et Louis Marie Mousset e la Lexicogrpahie Tamile*, Ph. D. Dissertation, Loyola College, Chennai, 2011, pp. 101-36.
91. CAOM, Inde 1831, *Rapport du Ministere de la Marine et des Colonies 26 Avril 1831, Proposition de Souscrire a Exemplaires d'un Dictionnaire Français-Malabar*.

92. CAOM, Inde n.d, *Catalogue des livres de l'Imprimerie de la Maison des missions Etrangères a Pondichéry, Circulaire addresse aux colonies et relative a une souscription a des ouvrages en langue tamoule edites par l'imprimerie de la mission Catholique en Inde*, 1863-1865.
93. CAOM, Inde 1863, *Envoie du catalogue des ouvrages en langue Tamoule edites par l'Imprimerie de la mission*, see the letter from the Governor of Pondicherry dated 18 February 1863 to the Minister of the Marine and Colonies in Paris.
94. Ibid.
95. Sonia N. Das, *Linguistic Rivalries: Tamil Migrants and Anglo-French Conflicts*, Oxford, 2016, p. 106.

Dictionnaire
tamoul, et
Français.

Préface.

8.1. The preface page from Tamil-French dictionary of 1724 by Louis-Noel de Bourzés (Courtesy: Bibliothèque Nationale, Paris).

Dictionnaire
Tamoul-François pour expliquer
le Tamoul Vulgaire
avec
une Préface
ou sont plusieurs Regles qu'il faut
sçavoir avant de faire usage du
Dictionaire
Composé par le Reverend Pere
Constant Ioseph Beschi de la
Compagnie de Iesus
Missionaire de l'Inde
l'an 1744
ad
Majorem Dei Gloriam, Virginis Matris
Mariæ Decus, et Religionis Christianæ
Catholicæ Romanæ Incrementum. Amen

8.2. The title page of Costanzo Giuseppe Beschi's Tamil-French dictionary of 1744 (Courtesy: British Library, London).

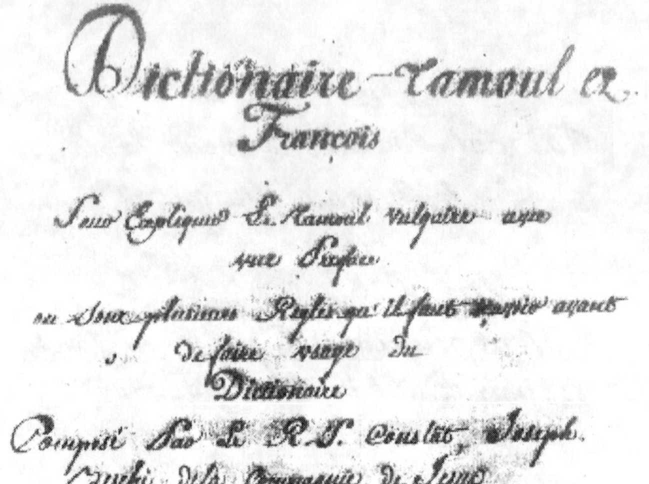
Dictionaire Tamoul et
François

Dictionaire
Compost Par Le R. P. Constant Joseph
Beschi de la Compagnie de Jesus

1806

Ad majorem Dei Gloriam, Virginis matris
Mariæ decus & Religionis Christianæ
Catholicæ Romanæ Incrementum
amen.

8.3. *Tamil-French Dictionary of Costanzo Giuseppe Beschi in 1744, a copy made in 1806 (Courtesy: French Institute Library, Pondicherry).*

DICTIONARIUM
LATINO-GALLICO-TAMULICUM.

DICTIONNAIRE
LATIN-FRANÇAIS-TAMOUL.

இலத்தின்-பிராஞ்சு-தமிழ்
அகராதி.

AUCTORIBUS DUOBUS MISSIONARIIS APOSTOLICIS
CONGREGATIONIS MISSIONUM
AD EXTEROS.

PUDICHERII,
E TYPOGRAPHIO MISSIONARIORUM APOSTOLICORUM
DICTÆ CONGREGATIONIS.

1846.

Facultate à S. Sede Apostolica
II. RR. Episcopo Drusiparensi V. A. concessâ.

8.4. The title page of Latin-French-Tamil dictionary printed in 1846 (Courtesy: Bibliothèque Asiatique des missions étrangères de Paris).

DICTIONNAIRE

TAMOUL-FRANÇAIS

PAR DEUX MISSIONNAIRES APOSTOLIQUES

DE LA CONGRÉGATION

DES MISSIONS-ÉTRANGÈRES.

அந்நியதேசத்துப்போதகர்சபையிலுள்ள

அப்போஸ்தொலிக்குக்குருக்களிருவர்

செய்தருளிய

தமிழ்ப்பிராஞ்சகராதி.

PONDICHÉRY,

IMPRIMERIE DES MISSIONNAIRES APOSTOLIQUES

DE LA DITE CONGRÉGATION

1855.

Avec l'approbation des Supérieurs.

8.5. The title page of Tamil-French dictionary volume one printed in 1855 (Courtesy: Bibliothèque Asiatique des missions étrangères de Paris).

DICTIONNAIRE

TAMOUL-FRANÇAIS,

PAR DEUX MISSIONNAIRES APOSTOLIQUES

DE LA CONGRÉGATION

DES MISSIONS-ÉTRANGÈRES.

அந்நியதேசத்துப்போதகர்சபையிலுள்ள

அப்போஸ்தொலிக்குக்குருக்களிலிருவர்

செய்தருளிய

தமிழ்ப்பிராஞ்சகராதி.

PONDICHÉRY,

IMPRIMERIE DES MISSIONNAIRES APOSTOLIQUES

DE LA DITE CONGRÉGATION.

1855.

Avec l'approbation des Supérieurs.

8.6. The title page of Tamil-French dictionary volume one printed with a different logo in 1855 (Courtesy: Bibliothèque Nationale, Paris).

DICTIONNAIRE

TAMOUL-FRANÇAIS,

PAR DEUX MISSIONNAIRES APOSTOLIQUES

DE LA CONGRÉGATION

DES MISSIONS-ÉTRANGÈRES,

TOME SECOND.

அந்நியதேசத்துப்போதகர் சபையிலுள்ள

அப்போஸ்தோலிக்குக் குருக்களி லிருவர்

செய்தருளிய

தமிழ்பிராஞ்சகராதியின்

உ-ம் பங்கு.

PONDICHÉRY,

IMPRIMERIE DES MISSIONNAIRES APOSTOLIQUES

DE LA DITE CONGRÉGATION,

1862.

Avec l'approbation des Supérieurs.

8.7. The title page of Tamil-French dictionary volume two printed in 1862 (Courtesy: Bibliothèque Asiatique des missions étrangères de Paris).

DICTIONNAIRE

FRANÇAIS-TAMOUL

PAR DEUX MISSIONNAIRES APOSTOLIQUES

DE LA CONGRÉGATION

DES MISSIONS - ÉTRANGÈRES.

...த்துப்போதகர்சபையிலுள்ள

...ப்போஸ்தொலிக்குருக்களிருவர்

செய்தருளிய

...ஞ்சுதமிழகராதி.

PONDICHÉRY,

IMPRIMERIE DES MISSIONNAIRES APOSTOLIQUES

DE LA DITE CONGRÉGATION

1866

Avec l'approbation des Supérieurs.

8.8. The title page of French-Tamil dictionary printed in 1866 (Courtesy: Bibliothèque Nationale, Paris).

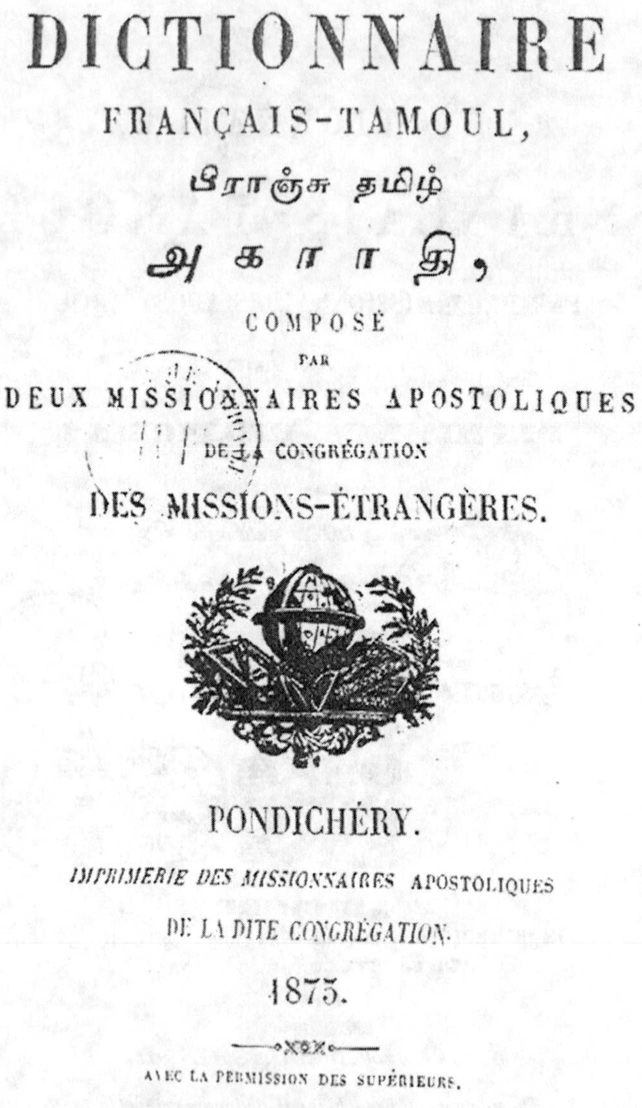

DICTIONNAIRE

FRANÇAIS-TAMOUL,

பிராஞ்சு தமிழ்

அகராதி,

COMPOSÉ

PAR

DEUX MISSIONNAIRES APOSTOLIQUES

DE LA CONGRÉGATION

DES MISSIONS-ÉTRANGÈRES.

PONDICHÉRY.

IMPRIMERIE DES MISSIONNAIRES APOSTOLIQUES

DE LA DITE CONGRÉGATION.

1875.

AVEC LA PERMISSION DES SUPÉRIEURS.

8.9. The title page of French-Tamil dictionary printed in 1875 (Courtesy: Bibliothèque Nationale, Paris).

9

Conclusion

Europe's connections to Tamil coast date back to the first millennium CE, when traders came from the Mediterranean and Red Sea regions to buy spices and aromatics. With the circulation of commodities, commerce had only developed and it seems to have been at its zenith between 50 BC and AD 50. With the fall of the Roman Empire the overseas trade also declined and it came to be stopped in the third century AD. After a very long interval contacts and relations had resumed in the early modern age with the Portuguese direct commercial and missionary expansion. The rendition of Christian prayer books, catechism and other literary texts in Tamil in the sixteenth century began and it led to the creation of unique linkages in language and grammar. It developed network across the European continent in the seventeenth century by triggering the foundation of new intellectual exchange and the publication of a wide array of works in Tamil language, grammar and dictionary writings. Tamil thus began to march towards global and interestingly this book unravels the missionary dialogue, discourse and difference.

The first of the proposition relates to the study of Tamil language and understanding of the grammar. We find each of the European missionary, Catholic or Protestant, found the Tamil grammatical works previously written in various European languages deficient in some respects and they began to improve upon them. Constructing the chapters on orthography and etymology they have followed more the order of the European grammars than that of the Tamil language and they judged it expedient to introduce the student at once to the western manner of treating the subject. They have noticed the difference between the Tamil and the European languages and gave various other differences between the European and Tamil idioms. Europeans have failed in giving us pure Tamil they have mixed vulgarisms with grammatical niceties and left us in want of a regularly digested syntax.

No other learned scholar like Constanzo Giuseppe Beschi had composed four grammatical works such as kodunthamizh in Latin, senthamizh in Latin, Thonnul Vizhlakkam in Tamil and Clavis in Latin. Beschi steered between the high and vulgar Tamil, explained well without avoiding the intricacies of the former and the barbarism of the latter. Europeans wrote about Tamil grammars in various European languages by using the grammar books and the works of literature such as the poems of Thiruvalluvar. They were aware that from the poetical writings, Tamil style and diction as used in common life could not be learnt, because they contained so many old Tamil words, which were not at all in use then and so many terminations and modes different from those in common conversation that it was not at all intelligible except after hearing the explanation from the learned men.

It may be mentioned that the several grammars composed by the Europeans have certain things in common. First they have been classified into mainly three parts. The first one dealt with letters, their sounds and pronunciation system, knowing several types of words and their usage and the conjugation of verbs. The second part dealt with word formation and how to employ them and the third part dealt with yapppilakkanam (prosody). Again these authors of the grammars pointed out the differences between Tamil and European languages grammar. They admit that pronunciation of certain letters is hard and difficult, but through rigorous exercise it would become easy. Pronunciation of certain words needs to be heard well from others while speaking. Tamil can be learnt through the grammar books prepared by them. It would take at least two or three years to learn to speak Tamil. More attention needs to be paid for learning senthamizh. They mentioned that Tamil language pronunciation varied according to different areas.

It is true that each author of the dictionary had made his own contribution to the dictionary-making. There has been continuity and discontinuity between them because they had different European language background. A mixture of Tamil and European methods emerged in dictionary-making in alphabetic format by different lexicographers. Antão Proença rejected poetic terms. Tamil alphabetical order for head-words was followed by de Louis-Noel de Bourzés. Most of these dictionaries were written for the use of the Europeans. The dictionary of Antão Proença is more closely linked to the spoken Tamil than the standard of today's prose which is based mainly on classical commentator's prose and purist reform efforts which excluded spoken elements and were not based on an analysis of modern language corpora. It must be said that spoken and literary forms of Tamil are two sides of the same coin and one without the other, either spoken or written Tamil could not have flourished through the ages. The dictionaries of the missionaries mainly compiled on spoken languages help us to know the non-poetic vocabulary of Tamil apart from inscriptions which were still free from European influence.

However, the western influence had impact in Tamil dictionaries. What was available in *nikandu* written in prose came to be read easily with separate words in format. Words came in alphabetical order reaching beyond the second letter till the last letter was introduced. Literary words and common words were compiled together and the wealth of words was shown in full form. Spoken words of ordinary people and those words used and coined by missionaries for the spread of Christianity were put together and found in dictionaries. Root words of the verbs were shown separately identified and from those words, other verbs used were given as examples.

After many years of learning Tamil, most of the missionaries attempted to add more words and increased the vocabulary and so composed dictionaries in several languages of Europe. Up-to-date dictionaries were desired and felt as a must for both the use of Tamils and Europeans and therefore many missionaries continued to undertake the dictionary work compilation. Each one contemplated to compile dictionaries in own way, style and method with their ability and skill in various European languages. As a result, the linguistic research on bi-lingual and tri-lingual developed compiling into one single dictionary which was really a welcome feature.

Thanks to the lexicographical programme of the missionaries, we find new domains of Tamil language use such as journalism and scientific discourse being developed towards future. Portuguese language though was a medium of instruction and interaction among the missionaries, for long, Tamil yet remained a major language and an essential means of access to the missionaries. In due course of time, words that existed in the bi-lingual dictionaries paved easy way for preparing the tri-lingual dictionaries. These dictionaries serve even today as treasure trove of information because they alone contain the database with inter-lingual features where many of the curious mixture of non-existing or obsolete Tamil and European words appear.

European missionaries not only compared Tamil with Greek and Latin but even with their own tongue in Europe. They pointed out that a general defect of Tamil was that though abundant in nouns it was very poor in verbs. They mentioned that Tamil language was still more barbarous in orthography and pronunciation because they neither separate the word nor the periods by full stops and commas and only in the beginning of their words do they used a vowel. They remarked that unless one learns this form of Tamil from his childhood, he can never pronounce the words with the same energy nor can their sound be properly written only with Portuguese letters, because there are numerous l's, t's and it is the same with the vowel A. It must be admitted that compilation of the dictionary by the missionaries was owing to the religious zeal and needs.

Many new developments had taken place and we find that the alphabetical arrangement of the words which were earlier limited in Tamil only to the second letter of the word came to be made from the first letter to the final letter i.e., the whole word, for dictionary entry construction. Easy words and difficult words were thus combined into one and thus Tamil words came in full shape. Missionaries who came and preached the Gospel to Tamils introduced all the colloquial words that were found in spoken language in their dictionaries. They separately identified and showed the root words of the verbs and showed the difference from them against the other verbs. They also showed the European words and the Tamil equivalent words in dictionaries. As learners of Tamil, European compilers also began to act as teachers of their own mother-tongue to Tamil speakers. Missionaries had their own well educated background, partly in the language, classics, rhetoric, grammar, and so they introduced novel purposes to which innovation into bilingual and trilingual dictionary construction was inevitable. Then came with their familiarity with the process of printing press, they had a considerable effect on the structural development of dictionary compilation in Tamil.

To estimate their work of linguistic scholarship at its true wisdom, one has to bear in mind the time when it was undertaken and the circumstances under which they achieved. The Portuguese were the pioneers in the fullest sense and they had to collect themselves, and copy them at a time in which none of the technical aids we have become dependent on, were at their disposal. The inconsistency in matter of Tamil and Portuguese spelling in the manuscripts is a well known fact and that was the order of the day.

It may be said that the dictionary-making was a very difficult work and exercise. Today Portuguese language is not familiar to many Tamil scholars hence they could not understand the glory of the works of the Portuguese. These missionaries had freed the Tamils from reading the *nikandus* composed in the poetry form and made them to learn the language easily. They had been the pioneers in making dictionary on the Tamil Coast and it served as a new manual for later scholar missionaries of Holland, Germany, France and England and for growth of a new discipline of linguistics.

It is gleaned that printed dictionaries contained a lot of words, increasing with head-words with an extensive vocabulary based firmly on the language of everyday affairs and the then present literature. It also incorporated many forms from the classical tradition which have entered the language of the people. It is primarily designed for French and Tamil speaking users and the dictionaries have served also to those who wish to improve their understanding of French language and Tamil language. The compilation of the dictionary, the purpose and scope of the dictionary and the selection of entries have differed from compilers to compilers and the presentation had some thought on learning Tamil and also learning French. The practical value had been its usefulness from linguistic view point. Missionaries like L.S. Dupuis and Mousset who learnt Tamil grammar had mainly composed bi-lingual dictionaries (both Tamil-French and French-Tamil). However, Dupuis and Mousset were the pioneers to compile and print dictionaries for the use of the locals, native-seminarians and overseas Tamils in the French colonies around the world.

It may be said that there is no end to dictionary work. A dictionary prepared for a language should be revised from time to time and it was the policy the missionaries followed. They felt that a new dictionary of Tamil language should be created and printed and this was the other idea that had struck them too. Further, researches into the dictionary-making work were also carried out by them. Printing and reprinting had also simultaneously happened. The missionaries in the beginning had difficulty in adding words particularly of Hindu origin. In the manuscript of Antão Proença word like Ramayanam were though present it was found missing in the printed dictionary at Ambalakaadu in 1679 for the inquisition and the Jesuit censor board authorities did not approve. Over a period of time the policy had changed and we find Hindu religious terms in the dictionary of Louis Savinien Dupuis printed at Pondicherry under the auspices of the MEP missionaries in 1855. Missionaries also did deletion of words, gave the double, triple and other meanings of a single word.

The missionary interest in Tamil language and literature, their deep involvement in Tamil studies/Tamilology was necessarily welcomed by their superiors at home who expected them to convert people rather than study the Hindu cultural traditions. As a result there was a tussle between Roberto Nobili and Goncalo Fernandez at Madurai in Tamil country. This also happened in the case of Ziegenbalg in Tranquebar and authorities at Halle in Germany who said that he wrote much about the heathen non-sense which they were worried about and they condemned. I have not come cross any single evidence showing that the missionary authorities in Europe took the fellow missionaries in Tamil country taking them to task for having devoted much time to scholarly work rather than for the apostolic/pastoral work. In fact learning Tamil language, grammar and dictionary-making was the work that sustained their spiritual energy under adverse circumstances. It is learnt that the European missionaries approached Tamil with the same reverence which they had shown when studying classical languages like Latin, Greek and Hebrew during the course of their priestly formation.

A drive for learning the Tamil grammar and composing dictionaries of bilingual and trilingual were to be considered the hall mark achievements of considerable importance, elevating Tamil by the missionaries from the local to the global and its spread across the oceans. The importance of Tamil was widely accepted and it was used as a medium of communication in Europe. Words given in the dictionaries in the Tamil-/Portuguese/French/ English section were more than the Portuguese/French/English-Tamil section and it was certainly enough for everyday use. Condensation was essential because all the information could not be provided in the dictionary. Maximum information was aimed and this has been done by the use of brackets and punctuation in the printed dictionaries. No attempts had been made by any European to compile only a Tamil-Dutch dictionary and a Dutch-Tamil dictionary. Bilingual constructions were necessary prerequisites for developing a catechetical literature for the direct use of the Christians. Scientific classification with the complete alphabetical arrangement of words in lexicography was the special feature of dictionaries compiled by Europeans. They had to create a dictionary with both forms of spoken and written Tamil and had to adopt the two styles in making of the work. Print language that appeared paved way for breaking the scribal culture of manuscripts. Missionary writings began to mirror the status of various speaker groups and caste groups in Tamil society. Modernization and democratization of Tamil developed by the Christian missionaries ushered Tamil into a new age of reading and writing with printed books. The work of the missionaries had crossed their traditions in a way that they had with a common European background, especially a Latin-based linguistic education. They shared immensely their knowledge in shaping Tamil language with linguistic data typologically of different languages from around the world in the early modern age.

Select Bibliography

Primary Sources

I. Manuscripts

Belgium

1. Bibliothèque Royale Albertine, Brussles

   MS no. 21868; no. 21881

Denmark

2. Det Kongelige Bibliotek, Kobenhavn

   MS NY. Kgl. Saml. 154, Christoph Theodosius Walther, Synonyma Ebraeo-Tamulica. Anno MDCCXL.

   MS NY. Kgl. Saml. 589C, Christoph Theodosius Walther, Bibliotheca Tamulica Consistens in Recesione Librorum Nostrorum, Mscr.torum, ad Cognoscendum et Linguam & Res Tamulicas Inseruientium, Anno MDCCXXXI.

   MS Rostg.40 2

   Oriental Department: Cod Tamul 30 (UB). Cod Tamul 31 (UB)

3. Rigsarkivet, Kobenhavn

   Asiatisk Kompagnis, MS 1450b, Vulgaris Tamulicae linguae dictionarium Tamulico-Latium additis in prefationae aliquot regulis necessario praelegendi, 1744.

   Ledr. 389.2

   Missionskollegiet og Direktionen for Vajsenhuset, 9G, Memoria for new missionaries from Christoph Samuel John and Johann Peter Rottler from Tranquebar dated 20 January 1784.

France

4. Archives de la société des missions étrangères de Paris

   MSS Correspondance Pondichéry-Paris, Lettres, vol. 997, 999, 1000, 1001, 1002.

Manuscrit Original

vol. 1062, Dictionnaire Tamoul–Français abrege de celui du R.P.Beschi

vol. 1091, Dictionnaire Tamoul–Français, Louis Savinien Dupuis

5. Archives Françaises de la Compagnie de Jésus, Vanves, Hauts-de-Seine

Fonds Brotier, vol. 86

MSS Index 3

6. Bibliothèque Nationale, Paris, Rue Richelieu

MSS Indien

no. 208-209. Dictionnaire Tamoul-Français, par Louis-Noel de Bourzés

no. 210. Dictionnaire Tamoul-Français, AD.1760

no. 211. Dictionnaire Malabare-Français

no. 212. Dictionnaire Malabare-Français

no. 213. Dictionnaire Tamoul et Français avec des notes grammaticales par Louis-Noel de Bourzés

no. 214. Dictionnaire Tamoul-Français par Louis-Noel de Bourzés

no. 215. Dictionaire Tamoul-Français pour expliquer le tamoul vulgaire avec une preface ou sont plusieurs regles qu'i l faut scavoir avant de faire usage du dictionnaire, compose par le Reverend Pere Constant Joseph Beschi de la Compagnie de Jesus missionaire de l'Inde l'an 1744 ad majorem dei gloriam, virginis matris maria deus et religionis Christiana Catholica Romana Incrementum

no. 216. Petit dictionnaire Français Tamoul compose par le Reverend Pere Constat Beschi, missionaire de la Compagnie de Iesu. Dans le Madure avec l'aide du Tamoul suffisant pour pouvoir s'enoncer sur tout

no. 217. Dictionnaire Tamoul-Français, precede d'un syllabaire, AD 1750

no. 219. Dictionnaire Français-Tamoul suivi de la langue malabare vulgaire

no. 220. Dictionnaire Français et Tamoul

no. 221. Arte Tamulica, Antão de Proença

no. 222. Arte Tamulica, Antão de Proença

MSS Nouvelles acquisitions françaises, no. 2627

MSS A. 2487/1

Germany

7. Archiv der Franckesche Stiftungen, Halle

   MS HA 61:1

   MS Letters: MB 75/76

   MSS M1 B2:7; M1 B2:26; M1 B26:29; M1 B73: 10(A); M1 C1:66, M1 C7:2, M1 H:12, MSS M2 B6:48; M2 B6:49; M2 B7:14b

8. Staats-und Universitatsbibliotheek Hamburg Carl von Ossietzy, Hamburg

   Cod. Orient 283, fl. 70, Arte Tamul Sive Instituto Gramatica Linguae Malabaricae Sum Philippi Baldej. UDM in Regno Jaffnapatam, 1659, Philippi Baldeyusu

India

9. Archives of the Gurukul Lutheran Theological College and Research Institute, Chennai

   Microfilms, Accession no. 578, Roll no. 51.

10. Government Oriental Manuscripts Library, Chennai

    MSS D8; D9

11. Historical Archives of Goa, Panaji

    MSS Monções do Reino, livro 177A, 182, 189, 191D, 192B, 193A, 195B, 195C

12. Institut Français de Pondichéry Library, Puducherry

    MSS DICT-TA, 13.1. Dictionaire Tamoul a Francois pour expliquee le tamoul vulgaire avec une preface ou souz plusieurs regles qui l'faut scavoir avant de faire usage du dictionaire compose par le R. P Constat Joseph Beschi de la compagnie de Jesus missionaire de de S.inde l'an (1744)- 1806 ad majorem dei gloriam, virginis matris maria decus et religionis Christiana catholica romana Incrementum Amen.

13. Madurai Province Jesuit Archives, Shenbaganur

    MS Tamil no. 3, 5 and 9, Vulgaris Tamulicae-Linguae Dictionarium, pars secunda in qua Lusitanis vocibus Latinae explicatis vulgaris idiomatis Tamulica verba adjunguntur

    no. 4, Constanzo Giuseppe Beschi, Vulgaris Tamulicae Linguae Dictionarium Tamulico-Latinum Additis in Praefationae Aliquot Regulis Necessario Praelegendis, 1744.

14. State Central Library, Goa.

MS no. 11. Chathuragaraathi: Quadruple dictionnaire Tamoul de Beschi, Contenant le Dictionaire des Mots, Celui des Synonymes, Celui des Classifications et Celui des Rimes, Anno 1741

no. 12. Padre Constantio Josepho Beschio, Dictionarium TamulicoLatinum, 1744, Part one of the dictionary, Tamil/Latin; Part two of the Dictionary, Latin/Tamil.

no. 15. Padre Francisco de Noronha, Significação de todas palavers no confessionario po mayor commoidade dos que começão a lingoa, Anno 1890.

no. 16. Padre Balthasar da Costa, Arte Tamulica

no. 32, Padre Balthasar Esteves da Cruz, Vocabulario Lusitano Tamulico, 1738

no. 33. Anónimo, Vocabulario lusitano-tamulico, século XVII

no. 34, Padre Antão de Proença, Vocabulário Tamulico-Luzitano Precedido de Uma Arte Abreviada em Tamul e Seguido de Termos Poéticos, Anno 1670

no. 35. Padre Anrique Anriquez, Vocabulario Lusitano-Tamulico Seguido d'uma Arte Tamulica, século XVI

no 36. Anónimo, Vocabulario lusitano-tamulico, século XVII

no. 37. Vocabulario thamulico lusitano para uso dos missionarios da Companhia de IESV. Composto, e augmentado pello Padre Domingos Madeyra da mesma Companhia, missionario na Missão de Madurey, Anno 1750

no. 38. Anónimo, Vocabulario tamulico-lusitano (in two columns) século XVIII

no. 39b. Padre Domingos Madeyra, Vocabulario thamulico lusitano para uso dos missionarios da Companhia de IESV. Composto, e augmentado pello P.…da mesma Companhia, missionario na Missão de Madurey, Anno 1750

no. 49 (fls.1-58) Arte Tamulica composta pello Padre Balthazar da Costa da Comp.a.de Iesu

no. 49 (fls.59 onwards) Vocabulario Lusitano-Tamulico pello Padre Balthazar da Costa da Comp.a.de Iesu

Italy

15. Archivum Romanum Societatis Iesu, Roma
    MS Goa vol. 8, 48, 54a
16. Biblioteca Apostolica Vaticana, Città del Vaticano.

Bon Compagai Section E6, MS Lettre di Roberto Nobili, Missionaria Gesuita nel Madure nellee Indie orientali dal anno 1606.

Borgiano Indiano 12

17. Bibliotheca Nazionale Centrale di Roma, Vittorio Emmanule II

MSS Varia 53 (598).

Portugal

18. Arquivo Historico Ultramarino, Lisboã

Documentos Avulsos, India Maço, 187 MS Reservado no. 1

19. Biblioteca da Ajuda, Lisboã

MSS Miscellanea Ultramarina: Codice 51-VI-30.

20. Biblioteca Municipal, O Porto

MS 804.

21. Biblioteca Nacional de Lisboã

MSS Reservados, Fundo Geral, Arte Malauar, no. 3141

22. Biblioteca de Sociedade de Geografia de Lisboa

MSS Reservados 146 (A) 14, fls.1-144v. Vocabulario Tamulico em Lusitano, Perture ao R.me Pe de Noronha Congragacao do Oratorio de cidade de Goa, Missionario Ilha de Ceylao, F.P. Pejalee 8 November 1838.

146(A) 16, Vocabulario ordenado para os que se applicaco ao Tamil, e Chingala comecado pello Portuguez, proseguido pello Tamil, e accabado pello Chingala pello Padre Euzebio do Rosario da Congreagação do Oratorio de Goa anno de 1772

146-A-17, Vocabulario Portuguese, Tamul e Chingala, copeado por hum certo Missionario de Ceylão, Pe. de Congreagação do Oratorio de S. Philippe Neri et Goa o qual foi composto por Padre Jacome Gonsalves da mesma Congreagação e superior que foi et mesma missão et Ceylão.

23. Institutio Arquivo Nacionais/Torre do Tombo, Lisboã

Documentos Remetidos da India, Livro 44

United Kingdom

24. British Library, London

Additional Manuscripts, no. 26110

European Manuscripts, E121. Dictionnaire et grammaire francois tamoul

Oriental Manuscripts, no. 1308, 13044. 13586, 14172

Sloane Manuscripts, no. 3003

25. John Rylands University Library, Manchester

Tamil manuscripts, no. 6, 7 and 11

United States of America

26. St. John's Abbey and University College Ville, Minnesota

A Arte da Lingua Malabar, microfilm no. 790

II. Printed Documents, Contemporary Chronicles, Tracts and Accounts

In English

A Grammar of the Common Dialect of the Tamulian Language Called Kodunthamizh Composed by R.F. Const. Joseph Beschi, Jesuit Missionary after a Study and Practice of Thirty Years, Translated by Christopher Henry Horst, Madras, Printed at Vepery, 1806.

A Grammar of the Common Dialect of the Tamul Language called Kodunthamizh; Composed for the use of the missionaries of the Society of Jesus [...], translated from the original Latin by George William Mahon, Printed by Reuban Twigg at the Christian Knowledge Society's Press, Vepery/Madras, 1848.

Babington, Benjamin Guy., *A Grammar of the High Dialect of the Tamil Language Termed Shen-Tamil to which is added an Introduction to Tamil Poetry: Translated from the Original Latin by Benjamin Guy Babington*, Printed at the College Press, Madras, 1822.

Fabricius, Johann Philipp., *A Grammar for Learning the Principles of the Malabar Language, Properly Called Tamul or the Tamulian Language, by the English Missionaries of Madras*, Printed at Wepery near Madras in the year MDCCLXXVIII.

Rhenius, Carl Theophilus Ewald., *A Grammar of the Tamil Language with an Appendix*, Church Mission Press, Madras, 1836.

The Report of the Sub-committee of the Madras Literary Society and Auxiliary of the Royal Asiatic Society, on Writing Indian Words in Roman Characters, Fort St. George Gazette Press, Madras, 1859.

In French

Catalogue des livres de l'Imprimerie de la Maison des missions etrangeres a Pondichery, Circulaire addresse aux colonies et relative a une

souscription a des ouvrages en langue tamoule edites par l'imprimerie de la mission Catholique en Inde, 1863-1865.

Catalogue des ouvrages en langue tamoule edites par l'Imprimerie de la mission, 1865.

Grammaire Française-Tamoule, ou les relges du Tamoul vulgaire, celles de la langue relevee et par celles de la poesie Tamoule se trouvent exposees au long par un provicaire Apostolique de la Congregation des Missions-Etrangères, Pondichéry, Imprimerie des Missionnaires Apostoliques de la dite Congregation, 1863.

Lettres édifiantes et curieuses, écrites des missions étrangères par quelques missionaires de la Compagnie de Jésus, 34 vols., Paris, 1702-1777.

Rapport du Ministere de la Marine et des Colonies 26 Avril 1831, Proposition de Souscrire a Exemplaires d'un Dictionnaire Français-Malabar.

In German

Boehme, Anton Wilhelm., *Propagation of the Gospel in the East: Being a further account of the progress made by some missionaries to Tranquebar upon the Coast of Coromandel, for the conversion of the Malabarians; of the methods by them taken for the effecting of this great work; of the obstructions they meet with in it, and of the proposals which they make, in order to promote it. Together with some observations relating to the Malabarian philosophy and divinity: and concerning their bramans, pantarms and poets*, London, Part II, second edition, 1711.

Der Königl. Dänischen Missionarien aus Ost-Indien eingesandter ausführlichen Berichten, Von dem Werck ihrs Ams unter den Heyden, angerichteten Schulen und Gemeinen, ereigneten Hindernissen und schweren Umstanden; Beschaffenheit des Malabarischen Heydenthums, gepflogenen brieflicher Correspondentz und mundlchen Unterredungen mit selbigen heyden, Teil 1–9, (Continuationen 1–108) Waiserihaus, Halle, 1710–1772.

Germann, Wilhelm., *Ziegenbalg und Plutschau: Die Grundungsjahre der Trankebarschen Mission: Ein Beitrag zur Geschite des Pietismus nach handscriftlichen Quellen und altesten Drucken*, 2 vols., Erlangen, 1868.

Knapp, Georg Christian., et.al., *Neuere Geschichte der evangelischen Missions-Anstalten zu Bekehrung der Heiden in Ostindien aus den eigenhändigen Aufsätzen und Briefen der Missionarien erausgegeben*, Teil 1–8, (Stück 1–95) Waisenhaus, Halle, 1770-8/ 1795-1848.

Lehmann, Arno., ed., *Alte Briefe aus Indien: Unveröffentlichte Briefe von Bartholomäus Ziegenbalg 1706–1719*, Berlin, 1957.

Niekamp, Johann Lucas., *Kurzgefaßte oder historischer Auszug der Evangelischen Missionsberichte aus Ost-Indien von dem Jahr 1705 bis zu Ende des Jahres 1736 mit zwey dazu nöthigen Landkarten und einer Vorrede Herrn Gotthilf August Franckens*, Halle, 1740.

Ziegenbalg, Bartholomäus., *An Account of the Religion and Government, Learning and Economy & etc of the Malabarians: Sent by the Danish Missionaries to their Correspondents in Europe (An rdEnglish Translation of the 3 Continuation of the Halle Reports Published in 1713), Printed and Sold by Joseph Downing in Bartholomew-Close near West Smithfield*, London, 1717.

__________, *Propagation of the Gospel in the East: Being a Collection of Letters from the Protestant Missionaries, and other worthy Persons in the East-Indies, & c. Relating to the Mission; the Means of Promoting it; and the Success it hath pleased GOD to give to the Endeavours used hitherto, for Propagating True Christianity among the Heathen in those Parts, but chiefly on the coast of Coromandel. With a map of the East-Indies, Part III, Published by the direction of the Society for Promoting Christian Knowledge, London, Printed and Sold by J. Downing*, in Bartholomew-Close near WestSmithfield, 1718.

In Latin

Beschi, C.J., *Grammatica Latino-Tamulica, ubi de Vulgari Tamulicae Linguae Idiomate, Kodunthamizh dicto, ad usum Missionariorum Soc. IESU, Auctore P. Constantino Iosepho Beschio, Ejusdem Societ. In regno madurensi Missionario AD MDCCXXVIII*, Trangambariae, Typis Missionis Danicae, MDCCXXXIX.

__________, *Grammatica Latino-Tamulica, in qua de Vulgari Tamulicae Linguae Idiomate, Kodunthamizh dicto, pusius tractatur, Auctore P. Constantino Iosepho Beschio, E Societate IESU, et in Regione Madurensi, apud Indes Orientale Missionario*, Apud Madraspatnam, E Typographeo Collegii, 1813.

__________, *Grammatica Latino-Tamulica in Qua De Vulgari Tamulicae Linguae Idiomate Kodun Thamzih Dicto, Fusisus Tractatur, Auctore P. Constantio-Josepho Beschio, Societatis Jesu, In Regione Madurensi, Apud Indos orientales, Missionario, NOVA EDITIO Cum notis et Compendio Grammaticae De Elegantiori Dialecto Sen Thamizh Dicata, An uno Missionario Apostolico Congragationis Missionum Ad Exteros,*

PUDICHERII, E Typographio Missionariorum Apostolicorum Dicte Congragationis, 1843.

__________, *Clavis Humaniorum Litterarum Sublimioris Tamulici Idiomatis, Auctroe R.P. Constantio Josepho Beschio Societatis Iesu, In Madurensis regno Missionario*, Tranquebar Evangelical Lutheran Mission Pesss, Tranquebar, 1876.

Conciones per Annum Concannice Compositae a patre Ignatio Archamone e Societatis Iesu, Goa, 1670.

Franco, Antonio., *Synopsis Annualium Societatis Jesu in Lusitania ab annis 1540 usque ad annum 1725*, Haeredum, Anno MDCCXXVI.

Ricci, Raphael., *Selectae Indicarum Epistolae nunc primum editae*, Florentiae, 1887.

Venuti, Filippo., *Dittionario volgare et latino nel quale si contiene come i vocaboli italiani si possono dire et esprimere latinamente per M. Filippo Venuti da Cortona con un dittionario delle voce latine similie à quello del Calepino co' numeri per ritrouar nel volgare la loro significatione aggiunto dal medesimo in questa nuova impressione*, In Venetia: appresso gli heredi di Luigi Valuassori & Gio. Domenico Micheli, 1582.

Walthero, Christophoro Theodosio., *Observationes Grammaticae quibus Linguae Tamulicae Idioma Vulgare inusum operariorum in messe Domini intergentes Malabares Dictas illustratur*, Trangambariae, MDCCXXXIX.

Ziegenbalg, B., *Grammatica Damulica, quæ per varia paradigmata, regulas & necessarium vocabulorum apparatum, viam brevissimam monstrat, qua Lingua Damulica seu Malabarica, quæ inter Indos Orientales in usu est, & hucusque in Europa incognita fuit, facile disci possit: in Usum eorum qui hoc tempore gentes illas ab idolatria ad cultum veri Dei, salutemque æternam Evangelio Christi perducere cupiunt: In itinere Europao, seu in nave Danica, concinnate a Bartholomaeus Ziegenbalg*, Halae Saxonum: Litteris & Impensis Orphanotrophei, MDCCXVI.

In Portuguese

Dal, Nikolaus., *Quarta parte da grammatica Portugueza, convem a saber hum vocabulario em portuguez e Malabar, que contem os nomes, verbos e adverbios de ambas as lingoas, que sao mais usados no trata cotidiano, e dos quaes os Vocabulos Malabares vao impressos com o carcater Tamul, e juntamente com letra Portugueza para o uso da escola Portugueza de Trangambar*, Trangambar, Na Officina da Real Missao de Dinmarca, Anno de MDCCXXXI.

Silva Rego, Antonio da., *Documentação Para a Historia das Missões do Padroado Portugues do Oriente*, 12 vols., Lisboã, 1947-58.

Stephen, S. Jeyaseela., ed., *Letters of the Portuguese Jesuits from Tamil Countryside, 1666-1688*, Pondicherry, 2001.

Wicki, Jose., and J. Gomes, *Documenta Indica* , 18 vols., Roma, 19481988.

In Tamil

Piranchu Ilakkana Noolsurukkam, Pondicherry, 1845.

Lathin Ilakkana Noorsurukkam, Pondicherry, 1845.

Ilakkana Surukkam, Pondicherry, 1845.

Ilakkana Noolaatharam, Pondicherry, 1849.

Ilakkana Surukkam, second edition, Pondicherry, 1860.

In Multi-lingual

Beschi, C.J., *Chaturagaraathi*, eds., Thandavaraya Mudaliar and Ramchandra Kavirayar, College Press, Madras, 1824.

__________, *Chathuragaraathi*, ed., Su. Innasi, Chennai, 2005.

Blin, Andre., *Agaraathi Thamzih: Dictionnaire Français-Tamoul et Tamoul-Français dedie a Moniseur Ministre de la Marine par A. Blin Lieutenant de Cypahis, Membre de la Société Asiatique de Paris*, Cher dendey Dupre Imp.lib. Rue Richelieu, 47 bis, a Paris.

Dupuis, Louis Savinien., *Dictionarium Latino-Gallico-Tamulicum: Dictionnaire Latin-Français-Tamoul, Latheen Thamizh Piranchu Agaraathi, Auctoribus duobus Missionariis Apostolicos Congregationis Missionum ad Exteros*, Puducherii, E Typographo Missionariorum Apostolicorum dictae Congregationis, 1846.

__________, *Vocabulaire Français-Tamoul, Compose par Deux Missionaires Apostoliques de la Congregation des Missions Etrangères et enrichi d'utiles renovis a leur dictionnaire Latin-Français-Tamoul*, Pondichéry, Imprimerie des Missionnaires Apostoliques, 1850.

__________, *Dictionnaire Tamoul-Français, par Deux Missionnaires Apostoliques de la Congregation des Missions Etrangères: Anniya Desaththu Pothagar Sabaiilulla Apostholikku Gurukkal Iruvar Seytharuliya Thamizh Piranchu Agaraathi*, Pondichéry, Imprimerie des Missionnaires Apostoliques de la dite Congregation, 1855.

__________, *Dictionnaire Tamoul-Français: Thamizh Piranchu Agaraathi Rendam Pangu, par Deux Missionnaires Apostoliques de la Congregation des Missions-Etrangères*, Pondichéry, Imprimerie des Missionnaires de la dite Congregation, 1862.

__________, *Dictionnaire Français-Tamoul: Piranchu Thamizh Agaraathi*, Pondichéry, Imprimerie des Missionnaires de la dite Congregation, 1866.

Fabricius, Johann Philipp, *A Malabar and English Dictionary, Wherein the Words and Phrases of the Tamilian Language, Commonly called by the Europeans the Malabar Language, are Explained in English by the English Missionaries of Madras*, Wepery, near Madras, 1779.

Fabricius, Johann Philipp and Johann Christian Breithaupt, *A Dictionary of the English and Malabar Languages, Printed and Completely done at Vepery*, near Madras, 1786.

Fabricius and Breithaupt, *A Malabar and English Dictionary Composed by the Rev. Messr. Fabricius and Breithaupt, English Missionaries at Madras and Published in the Year 1779; The Second Edition Revised and Corrected by the Rev. Mr. Poezold, Mr. William Simpson and the Malabar Catechists, Printed at Vepery in the Year 1809*, Madras, 1809.

Fabricius, Johann Philipp, *A Dictionary of the Tamil and English Languages by J. P. Rottler, Missionary of the Society for the Propagation of the Gospel in Foreign Parts*, vol. I, Part I, Madras, Printed at the Vepery Mission Press, 1834.

__________, *A Dictionary of the Tamil and English by the Late Reverend J. P. Rottler Missionary of the Society for the Propagation of the Gospel in Foreign Parts*, Vol. I, Part II, Madras, Printed at the Vepery Mission Press, 1836.

__________, *A Dictionary of the Tamil and English by the Late Reverend J. P. Rottler Missionary of the Society for the Propagation of the Gospel in Foreign Parts*, Part III, Revised by W. Taylor and T. Vencatachala Moodelly, Printed by K.W. Thrope, at the Vepery Mission Press of the Society for Promoting Christian Knowledge, 1839.

__________, *A Dictionary of the Tamil and English by the Late Reverend J. P. Rottler, Part IV, Revised by W. Taylor and T. Vencatachala Moodelly*, Madras, Printed at the Vepery Mission Press, 1841.

Thaninayagam, S. Xavier., *Antão de Proença's Tamil-Portuguese Dictionary, AD 1679*, Kuala Lumpur, 1966.

Vermeer Hans, J., and Angelika Morath, *The First European Tamil Grammar*, Heidelberg, 1982.

Vocabulario Tamulico com a significacam Portugueza composto pello p. Antam de Proenca da Companhaia de Jesu, Missionario da Missam de Madurey, com todas as licencas necessaries da Santa Inquizicao e dos superiors, na impressa Tamulica do provincia do Malabar, por

Ignacio Archamoni impressor della, Ambalacatta em 30 Julho 1679 annos.

III. Travelogues

Baldaeus, Philippus., *Nauwkeurige beschryvinge van Malabar en Choromandel, derhalver aangrenzende ryken, en het machtige eyland Ceylon. Nevens een omstandige en grondigh doorzochte ontdekking en wederlegginge van de afgoderye der Oost-Indische heydenen*, 3 vols, Amsterdam, 1672.

__________, *A True and Exact Description of the Most Celebrated East India Coasts of Malabar and Coromandel as well as of the Isle of Ceylon with their Adjacent Kingdom and Provinces*, III vols., London,1703, rpt., New Delhi, 2000.

Barbosa, Duarte., *The Book of Duarte Barbosa: An Account of the Countries Bordering of the Indian Ocean and their inhabitants*, 2 vols., ed., L.M. Dames, London, 1918–21, rpt., Delhi, 1989

Duperron, Anquetil., *Relation de Voyage en Preliminiaire a la Production du Zend-Avesta*, eds., Jean Deloche, Manonmani Filliozat and Pierre Sylvain Filliozat, *Voyage en Inde, 1754-1762*, Paris, 1997.

San Bartolomeo, Paolonio da ., *Voyage to the East Indies*, London, 1800.

Secondary Works

Articles

Chevillard, Jean-Luc, 'Beschi, grammairien du tamoul, et l'origine de la notion de verbe appellatif'. *Bulletin de L'Ecole française d'Extrême-Orient*, Tome 79, No. 1, 1992, pp. 77-88.

'Description of Two Manuscripts in the Library of Madras Literary Society. De Bourges' Dictionnaire Tamul-Francais (French Preface of De Bourges)', *Madras Journal of Literature and Science*, Third Series, 1864, pp. 111-15.

Duverdier, Gérald., 'L'Imprimerie tamoule de vepery et es deux dictionnaire de Fabricius', *Gutenber-Jahrbuch*, 1978, pp. 189-97.

Gaur, Albertine., 'Bartholomäus Ziegenbalg's Verzeichnis der Malabarischen Bucher', *Journal of the Royal Asiatic Society of Great Britain and Ireland*, no. 3-4, October, 1967, pp. 63-95.

Hein, Jeanne., 'Father Henriques' Grammar of Spoken Tamil', *Indian Church History Review*, Vol. 11, 1977, pp. 127-57.

James Gregory, 'The Terminology of Declension in Early Missionary Grammars of Tamil', in *Missionary Linguistics III/Linguistica Misionera III, Morphology and Syntax: Selected Papers from the Third and Fourth International Conferences on Missionary Linguistics, Hong Kong/Macau, 12-15 March 2005, Valladolid, 8-11 March 2006*, eds., Otto Zwartjes, Gregory James and Emilio Ridruejo, Amsterdam, 2007, pp.167–190.

__________, 'Aspects of the Structure of Entries in the Earliest Missionary Dictionary of Tamil', in *Missionary Linguistics IV/Linguistica Misionera IV, Lexicography: Selected Papers from the Fifth International Conference on Missionary Linguistics, Merdia, Yuctan, 14-17 March 2007,* eds., Otto Zwartjes, Ramon Arzápalo Marín and Thomas C. Smith-Stark, Amsterdam, 2009, pp. 273–302.

Knowlton, Edgar C., 'Antão de Proença's Vocabulario Tamulico Lusitano: Indo-Portuguese Elements', *Tamil Culture*, vol. XI, 1964, pp. 135-64.

__________, 'Portuguese-Tamil Linguistic Contacts', in *Proceedings of the First International-Seminar-Conference of Tamil Studies*, 1966, pp. 607-17.

Malten, Thomas., 'Antam de Proença's Tamil-Portuguese Dictionary, Printed in India in 1679', in Charles J. Borges and Helmut Feldmann, eds., *Goa and Portugal: Their Cultural Links*, New Delhi, 1997, pp. 271-79.

Muru, Cristina., 'Gaspar de Aguilar: A Banished Genius' in Anand Amaladass and Ines G. Zupanvo, eds., *Intercultural Encounters and the Jesuit Mission in South Asia, 16th -18th Centuries*, Bangalore, 2014, pp. 353-389.

Pillei, A. Muttusami., 'A Brief Sketch of the Life and Writings of Fr. J. Beschi or Veeramamuni', *Madras Journal of Literature and Science*, vol. 11, no.27, 1840, pp. 250-300.

Soler, Elena Losada., 'The Encounter of Languages: Reflection on the Language of the Other in Roteiro da Primeiro Viagem de Vasco da Gama' in Anthony Disney and Emily Booth, *Vasco da Gama and the Linking of Europe and Asia*, New Delhi, 2000, pp. 200-19.

Stephen, S. Jeyaseela., 'Indian Manuscripts Sent from Chandernagore and Pondicherry (AD 1714-1735): Role of the French East India Company and Jesuit Missionaries', *Journal of the Oriental Institute*, vol. 50, no. 1-4, September-June 2000-2001, pp. 245-248.

__________, 'Portuguese Contribution to Tamil Lexicography', Indrani Mukherjee and Sovon Sanyal, eds., *Reading "Culture" in Spanish & Luso-Brazilian Studies*, New Delhi, 2004, pp. 354-77, see, pp. 372-4.

Subramanian, P.R., 'The Metamorphosis of Tamil-English Lexicography: Dictionaries of Foreign Missionaries in the Period 1750-1900', in C.T. Indra and R. Rajagopalan, *Culture, Language & Modernity: English-Tamil in Colonial India, 1750-1900*, Abingdon, 2018, pp. 30-47.

Sundaram, T.P. Meenakshi., 'A Study of Baldaeus's Grammar', *Indian Linguistics*, vol. 1, 1958, pp. 7-19.

__________, 'Tamil Letters and Sounds of the Seventeenth Century', *Collected Papers of Prof. T. P. M, Sixty-first Birthday Commemoration Volume*, Annamalai Nagar, 1961, pp. 225-31.

Thaninayagam, Xavier., 'Tamil Manuscripts in European Libraries', *Tamil Culture*, vol. III. October 1954, pp. 210-88.

__________, 'Antão de Proença's Tamil-Portuguese Dictionary, 1679, An Introduction', *Tamil Culture*, vol. XI, no.2, April-June 1964, pp, 117-27.

Van Buiteen J.A.B., and P. C. Ganesh Sundaram, 'A 17th Century Dutch Grammar of Tamil', *Bulletin of the Deccan College Research Institute*, Poona, vol. XIV, no. 3. December 1952, pp. 168-82.

Vinson, Julien., 'Note Supplementaire sur les Oeuvres Tamoules du P. Beschi', *Revue de Linguistitue et de Philologie Comparée*, vol. XXXIII, 1900, pp. 198-200.

Wicki, Josef., 'Life of P. Henrique Henriques, S.J. (1520-1600)', *Studia Missionalia*, vol. XIII, 1963, pp. 157-160.

Zaleski, Michelle., 'Beyond Words: Missionary Grammars and the Construction of Language in Tamil Country', *Encounters between Jesuits and Protestants in Asia and the Americas*, Leiden, 2018, pp. 159-170.

Books

Anchukandam, Thomas., *The First Synod of Pondicherry 1844: A Study Based on Archival Sources*, Bangalore, 1994.

Babington, Benjamin Guy., *A Grammar of the High Dialect of the Tamil Language Termed Shen-Tamil to which is added an Introduction to Tamil Poetry: Translated from the Original Latin by Benjamin Guy Babington*, Printed at the College Press, Madras, 1822.

Besse, Leon., *La Mission du Maduré: Historique de Ses Pangous*, Trichnopoly, 1914.

__________, *A Grammar of High-Tamil*, St. Joseph's Industrial School Press, Trichinopoly, 1917.

__________, *Father Beschi: His Times and his Writings*, Trichnopoly, 1918.

__________, *Liste Alphabetiques des Missionaire du Carnatic de la Compagnie de Jesus au XIIIe Siecle*, Pondichéry, 1918.

Catalgo dos Livros, Opusculos e Manuscriptos Pertences a Bibliotheca Nacional de Nova Goa, Nova Goa, 1907.

Costelloe, M. Joseph., trs., *The Letters and Instructions of Francis Xavier*, Anand, 1993.

Dhamotharn, A., *Tamil Dictionaries-A Bibliography*, Wiesbaden, 1978.

Fenger, Johann Ferdinand, *History of the Tranquebar Mission: Worked out from Original Papers*, trs., Emil Francke, Tranquebar, 1863.

Hein Jeanne, and V.S. Rajam, *The Earliest Missionary Grammar of Tamil: Fr. Henriques' Arte da Lingua Malabar: Translation, History and Analysis*, Cambridge, Harvard University Press, 2013.

Hudson, Dennis D., *Protestant Origins in India: Tamil Evangelical Christians, 1706-1835*, Surrey, 2000.

Innasi, Su., *Thembavani Thiran*, Chennai, 1994.

James, Gregory., *Tamil Lexicography*, Tubingen, 1991.

__________, *Col Porul: A History of Tamil Dictionaries*, Chennai, 2000.

Jayadevan, V., *Tamil Akarathiiyal Valarchi Varalaru*, Madras, 1985.

Jeyaraj, Daniel., *Bartholomäus Ziegenbalg: The Father of Modern Protestant Mission, An Indian Assessment*, Delhi, 2006.

__________, *Tamil Language for Europeans: Ziegenbalg's Grammatica Damulica (1716): Translated from Latin and Tamil*, Wiesbaden, 2010.

Lafont, Jean-Marie., *Indika: Essays in Indo-French Relations, 1630-1976*, Delhi, 2000.

Launay, Adrien., *Histoire des Missions de l'Inde: Pondichéry, Maissour, Coimbatour*, 4 vols., Paris, 1895-1898.

Lehmann, Arno., *Es begann in Tranquebar: Die Geschichte der ersten evangelischen Kirche in Indien*, Berlin, 1955.

Meenakshisundaram, K., *The Contribution of European Scholars to Tamil*, Madras, 1974.

Mohanavelu, C.S., *German Tamilology: German Contributions to Tamil Language, Literature and Culture during the Period 1706-1945*, Madras, 1993.

Muru, Cristina., *Missionari Portoghesi in India nei Secoli XVe XVIIe: L' arte della lingua Tamil: Studio Comparati di alcuni Manoscritti*, Roma, 2010.

Norgaard, Anders., *Mission U.Obrigkeit: Die Danisch-Hallische Mission in Tranquebar, 1706-1845*, Gutersloh, 1988.

Pillai, Pakkiam., *Dictionarium Tamulico-Latinum*, South India Times Press, Tiruchirapalli, 1882.

Pillei, Muttusamai., *A Brief Sketch of the Life and Writings of Fr. C.J. Beschi*, Madras, 1840.

Pilley, Pakkiam., ed., *Vulgaris Tamulicae Linguae Dictionarium Tamulico-Latinum: Additis in Praefatione aliquot Regulis Necessario Praelegendis,* Typis South India Times Dictis, Trichnopoly, 1882.

Pingala Nikandu, South India Saiva Siddhanta Pathippu Kazhagam, Tirunelveli, Second edition, 1978.

Raghavacharyar Srinivasa., and G. Mackenzie Cobban, eds., *Thonnul Vilakkam: Mulamum Uraiyum*, Madras, 1891.

Rajamanickam, S., *The First Oriental Scholar*, Palayamkottai, 1972.

__________, *Viramamunivar: Thondum Pulamaiyum*, Madras, 1996.

Rivara, J. H. Cunha., *A Jurisdiçaõ Diocesana do Bispado de S. Thomé de Meliapor nas Possesções Francezas e Inglezas*, Nova Goa, 1867.

Sauliere, A., *His Star in the East* (Revised and reedited by S. Rajamanickam), Anand, 1995.

Schurhammer Georg., *Epistolae S. Francisci Xaverii alia que eius Scripta*, Roma, 1944.

__________, *St. Francis Xavier: His Life, His Times*, 2 vols., Rome, 1977.

Shanmuganar, Sundara., *Tamil Akarathi Kalai*, Pondicherry, 1965.

Stephen, S. Jeyaseela., *Portuguese in the Tamil Coast: Historical Explorations in Commerce and Culture, 1507-1749*, Pondicherry, 1998.

__________, *Muthu Vijaya Thiruvengadam Pillai Naatkurippu, 1794-1796*, Pondicherry, 1999.

__________, *Portuguese-Tamil Grammar: Modernization and Democratization of Tamil*, Pondicherry, 2001.

__________, *Letters of the Portuguese Jesuits from Tamil Countryside, 1666-1688*, Pondicherry, 2001.

__________, *The Diary of Rangappa Thiruvengadam Pillai, 17611768, Translated from Original Tamil*, Pondicherry, 2001.

__________, *Caste, Catholic Christianity and the Language of Conversion: Social Change and Cultural Translation in Tamil Country, 1519-1774*, New Delhi, 2008.

__________, *Expanding Portuguese Empire and the Tamil Economy: Sixteenth-Eighteenth Centuries*, New Delhi, 2009.

__________, *Oceanscapes: Tamil Textiles in the Early Modern World*, Delhi, 2014.

__________, *A Meeting of the Minds: European and Tamil Encounters in Modern Sciences, 1507-1857*, Primus, Delhi, 2016.

__________, *Louis Savinien Dupuis (1806-1874): A Biography of the MEP Missionary in Pondicherry*, Pondicherry, 2017.

Tholkaapiyam, South India Saiva Siddhanta Pathippu Kazhagam, Madras, 1982.

Veerasami, V., *Akarathikalai*, Madras, 1989.

Županov, Ines G., *Disputed Mission: Jesuit Experiments and Brahmanical Knowledge in Seventeenth Century India*, New Delhi, 1999.

__________, *Missionary Topics; The Catholic Frontiers in India, 16th-17th Centuries*, Ann Arbor, 2005.

Zwartjes, Otto., *Portuguese Missionary Grammars in Asia, Africa and Brazil, 1550-1800*, Amsterdam, 2011.

Theses and Dissertations

Gnaanapragsam, V.M., *Contribution of Fr. Beschi to Tamil*, Unpublished Ph.D. Dissertation, University of Madras, 1965.

Manickam, Michael., *On Describing a Non-Indo-European Language in Traditional Indo-European Grammatical Categories*, Unpublished Master's Thesis, Hartford Seminary Foundation, Hartford, 1968.

Raj, Gabriel Francis Xavier., *L'apport francais a la lexiocographie Tamile*, Jawaharlal Nehru University, Unpublished Ph.D Dissertation, 1995.

Roy, Navis Sybil Abarna, *La Contribution de Louis Savinien Dupuis et Louis Marie Mousset e la Lexicogrpahie Tamile*, Unpublished Ph.D. Dissertation, Loyola College, Chennai, 2011.

Index

V

W

X

Y

Z